PROGRAMMING AND COMPUTER TECHNIQUES IN EXPERIMENTAL PHYSICS

PROGRAMMIROVANIE I VYCHISLITEL'NAYA TEKHNIKA V FIZICHESKOM EKSPERIMENTE

ПРОГРАММИРОВАНИЕ И ВЫЧИСЛИТЕЛЬНАЯ ТЕХНИКА В ФИЗИЧЕСКОМ ЭКСПЕРИМЕНТЕ

The Lebedev Physics Institute Series

Editor: Academician D. V. Skobel'tsyn

Director, P. N. Lebedev Physics Institute, Academy of Sciences of the USSR

Volume 25	Optical Methods of Investigating Solid Bodies
Volume 26	Cosmic Rays
Volume 27	Research in Molecular Spectroscopy
Volume 28	Radio Telescopes
Volume 29	Quantum Field Theory and Hydrodynamics
Volume 30	Physical Optics
Volume 31	Quantum Electronics in Lasers and Masers
Volume 32	Plasma Physics
Volume 33	Studies of Nuclear Reactions
Volume 34	Photomesonic and Photonuclear Processes
Volume 35	Electronic and Vibrational Spectra of Molecules
Volume 36	Photodisintegration of Nuclei in the Giant Resonance Region
Volume 37	Electrical and Optical Properties of Semiconductors
Volume 38	Wideband Cruciform Radio Telescope Research
Volume 39	Optical Studies in Liquids and Solids
Volume 40	Experimental Physics: Methods and Apparatus
Volume 41	The Nucleon Compton Effect at Low and Medium Energies
Volume 42	Electronics in Experimental Physics
Volume 43	Nonlinear Optics
Volume 45	Programming and Computer Techniques in Experimental Physics

In preparation

Volume 44	Nuclear Physics and Interaction of Particles with Matter
Volume 46	Cosmic Rays and Interaction of High-Energy Particles
Volume 47	Radio Astronomy: Instruments and Observations
Volume 48	Surface Properties of Semiconductors and Dynamics of Ionic Crystals
Volume 49	Quantum Electronics and Paramagnetic Resonance
Volume 50	Electroluminescence

Proceedings (Trudy) of the P. N. Lebedev Physics Institute

Volume 45

PROGRAMMING AND COMPUTER TECHNIQUES IN EXPERIMENTAL PHYSICS

Edited by

Academician D. V. Skobel'tsyn

Director, P. N. Lebedev Physics Institute
Academy of Sciences of the USSR, Moscow

Translated from Russian

CONSULTANTS BUREAU
NEW YORK–LONDON
1970

The Russian text was published by Nauka Press in Moscow in 1969 for the Academy of Sciences of the USSR as Volume 45 of the Proceedings (Trudy) of the P. N. Lebedev Physics Institute. The present translation is published under an agreement with Mezhdunarodnaya Kniga, the Soviet book export agency.

Труды Ордена Ленина
Физического института им. П. Н. Лебедева
том 45

Library of Congress Catalog Card Number 71-118860
SBN 306-10843-7

© 1970 Consultants Bureau, New York
A Division of Plenum Publishing Corporation
227 West 17th Street, New York, N. Y. 10011

United Kingdom edition published by Consultants Bureau, London
A Division of Plenum Publishing Company, Ltd.
Donington House, 30 Norfolk Street, London, W.C. 2, England

All rights reserved

No part of this publication may be reproduced in any
form without written permission from the publisher

Printed in the United States of America

CONTENTS

The Calculation of Nucleon Cascades in Nuclei by the Monte-Carlo Method.
    R. A. Latypova 1

Beam Behavior in a Sector Cyclotron. A. T. Matachun, V. A. Gladyshev,
    L. N. Katsaurov, E. M. Moroz, and L. P. Nechaeva 5
    Introduction 5
    1. Harmonic Analysis of the Magnetic Field 5
    2. Determination of Equilibrium Orbits 6
    3. Orbit Stability, Betatron Oscillations 8
    4. Analysis of Vertical Motion 9
    5. The Calculation of Trajectories for Ion Drift in an Inhomogeneous
       Magnetic Field 11
    Literature Cited 11

Vassal (Automatic Memory Allocation System). U. G. Martin 13
    Introduction 13
    1. Input Language for the Vassal Translator 14
    2. Some Remarks on Programming in Vassal 17
    3. Translator Operation 18
    4. Rules for Pending Vassal Program 21
    Literature Cited 21

Determination of Reactions Cross Sections from Counter-Telescope Data.
    L. I. Slovokhotov 22
    Introduction 22
    Chapter I. Relation Between the Output and the Cross Section .. 24
    1. Formulation of Experiment. Definitions 24
    2. Calculation of the Cross Section from the Output without Correction
       for Multiple Scattering 24
    3. Multiple Scattering 29
    Chapter II. Calculation of the Output in the Case of the Compton Effect
    and the Photoproduction of Neutral Mesons on Hydrogen 38
    1. Definitions and Additional Restrictions 38
    2. Evaluation of the Multiple Integral 41
    3. Analysis of the Output Integrand 44
    4. Determination of the Range of Integration for the Entire Integrand 49
    Chapter III. Program for Calculations on the M-20 Computer 56
    1. Description of Program 56
    2. Program Checks 61
    3. Conclusions 62
    Literature Cited 62

CONTENTS

Typical and Atypical Failures of the General-Purpose Computer M-20, Methods of Localization and Elimination. V. V. Gavrilov, B. G. Minaev, and Yu. V. Stupin ... 63
   Introduction ... 63
  Chapter I. Failure Detection from the Control Console ... 63
    1. Tests and Test Problems ... 63
    2. Fault Analysis in the Execution of Individual Instructions ... 73
  Chapter II. Failure Detection by Technical Means ... 76
    1. Choice of Synchronization Method for the Localization of Intermittent Failures ... 76
    2. Utilization of Auxiliary Circuits for the Localization of Transient Failures ... 77
    3. Preventive Maintenance ... 79
  Chapter III. Typical Failures of the Basic Devices on M-20 ... 80
    1. Typical Failures of CM and Their Elimination ... 80
    2. Typical Failures in Backup Storage (Magnetic Drums and Tapes) and Their Control Circuits ... 81
    3. Typical Failures in the Arithmetic Unit ... 83
    4. Control Unit and Its Typical Failures ... 84
  Chapter IV. Examples of Atypical Failures in M-20 ... 85
    1. Failures in CM ... 86
    2. Failures in AU ... 88
    3. Failures in CU ... 89
    4. Failures in the Input-Output Devices ... 90
  Chapter V. Certain Questions of Improving the Machine During Its Exploitation ... 92
    1. Reinforcement of Weak and Most Heavily Loaded Circuits and Suppression of Oscillations ... 93
    2. Reduction of the Execution Time of Certain Operations ... 97
    3. Other Changes in the Circuits of M-20 ... 100
   Literature Cited ... 101
   Appendix ... 101

Some Problems in Analyzing the Dynamic Structure of an Object from the Steady-State Signal. L. I. Gudzenko ... 104
   1. Inverse Analysis Problems ... 104
   2. Statement of the Problem ... 106
   3. Method of Obtaining the Dynamic Operator ... 110
   4. Region of Application of Method ... 120
   5. Comments ... 126
   Literature Cited ... 126

THE CALCULATION OF NUCLEON CASCADES IN NUCLEI BY THE MONTE-CARLO METHOD

R. A. Latypova

Analytic calculations of nucleon cascades are difficult in principle and are not possible at the present time because some of the characteristics of a cascade, for example, the nucleon–nucleon cross sections, do not have an adequate analytic description. The Monte Carlo method allows us to perform an approximate calculation of a nucleon cascade and to determine the necessary nuclear characteristics without using analytic expressions.

Let us suppose that a beam of protons is incident on a nucleus. The energy and the initial direction of the protons are given. The point of entry, the mean free path for each particle, and the point of a possible collision with a nucleon of the nucleus are selected at random. In a collision, a particle may cease to exist (be absorbed), may scatter (i.e., receive a new direction and a new energy), or after several collisions it may leave the nucleus.

The types of collisions and the forms of the interaction between the incident particle and nuclear particles are also selected at random. Eight forms of interaction were taken into account: elastic nucleon–nucleon interactions, inelastic interactions with the formation of one or two π-mesons; in π-meson–nucleon collisions, absorption, scattering with charge exchange, and interactions leading to the production of one or two mesons. The relative probabilities of the various types of collisions, the scattering cross sections, and the angular distributions of created particles were taken the same as in [1]. The pair-correlation model was used to calculate π-meson absorption. Each created particle is followed in exactly the same manner. The result is a branching trajectory (description of the cascade). A branching trajectory is constructed for each incident proton. The collection of trajectories is used for the approximate determination of the nuclear characteristics of interest to us, namely, the energy and angular distributions, etc.

Each trajectory is constructed independently of the others, so that once it is obtained, it can be processed immediately and the results stored before the next trajectory is constructed. The number of references to the backing store (slow memory) required with this computational scheme is considerably less than would be the case if all trajectories had to be stored before processing could take place.

The calculation of nuclear cascades does not require the construction of an entire trajectory. A trajectory is processed by stages according to the following scheme: a single branch is processed first and all branch points on it are recorded; having reached the end of a branch, we return to the node immediately preceding the end point and destroy it; we then move along one of the recorded branches. If there are no branches at a given point, we move backwards one node. The computation ends when we have considered all of the branches of a given trajectory. This is the principal scheme of the Monte-Carlo method used for the calculation of a nuclear cascade. The following basic assumptions were used in the calculations:

1. Collisions of cascade nucleons were calculated relativistically in three-dimensional geometry.

2. Neutrons and protons were distinguished by charge and collision cross sections. Their masses, however, were assumed to be the same ($mc^2 = 938.85$ MeV). The mesons were also distinguished by their charge (+, −, 0), but their masses were assumed to be the same and equal to $m_\pi c^2 = 137$ MeV.

3. The density distribution of nuclear matter was described by a distribution function close to that obtained in experiments with fast electrons, but not identical with the latter, namely,

$$\rho(r) = \frac{\rho_1}{1 + e^{r-c \cdot z}}, \qquad r \leqslant R,$$
$$\rho(r) = 0, \qquad r > R,$$

where R is the radius of the nucleus defined by

$$\rho(R)/\rho(0) = 0.1.$$

The values of c and z were taken from [2].

4. The potential of the nucleon–nucleus interaction was taken in the form of a rectangular well of radius R. The depth of the well was taken to be the same as that used in [1].

5. The total cross sections (σ_{Av}) for the collisions between incident nucleons and π-mesons and nucleons in the nucleus were obtained by an averaging of nucleon–nucleon and meson–nucleon cross sections over the various types of collisions. The nucleon–nucleon cross sections were taken from [1], the meson–nucleon cross sections were taken from [3].

6. The range of a nucleon in nuclear matter was calculated with nonuniformities of the nuclear density distribution taken into account. The path length x_N corresponding to a random number N was calculated from the formula

$$\int_0^{x_N} \rho(t)\,dt = -\frac{1}{\sigma_{av}} \ln \frac{\sqrt{(N_0 - N)(N_0 - N + 1)}}{N_0},$$

where $N_0 = 100$ is the number of equiprobable subdivisions of the path length and $\rho(t)$ and σ_{Av} are defined in 3 and 5 above. This equation was solved numerically.

7. After the coordinates r, θ, φ of the collision point have been determined, the condition r < R was checked. If r ≥ R, we assumed that the particle did not suffer any collisions inside the nucleus. When r < R, we assumed that a collision with one of the nucleons in the nucleus had taken place.

8. Two types of momentum distributions for the nucleons in the nucleus were considered – a Fermi distribution and a truncated Gaussian distribution with the same value of the rms momentum (18 MeV). The cutoff energy for the Gaussian distribution was 49 MeV.

9. Eight types of collisions were taken into account.

10. We checked whether a collision is allowed by the Pauli principle after the results of the collisions had been calculated.

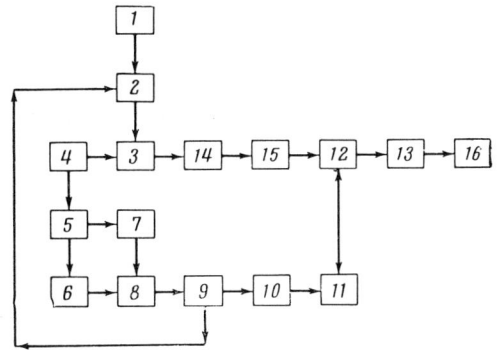

Fig. 1. Flow diagram of the computation.

11. A nucleon was assumed to be absorbed after a collision if its kinetic energy was less than the average depth of the potential well (for protons and neutrons). A particle that had not suffered any collisions inside the nucleus was considered to have escaped from the nucleus if its energy was greater than the sum of the energies of the centrifugal potential and the nuclear potential.

The momentum of the particle outside a nucleus was obtained with refraction at the nuclear boundary taken into account. A particle whose kinetic energy was insufficient to overcome the nuclear-force potential and the centrifugal potential was assumed to be reflected. The angle of reflection was taken equal to the angle of incidence.

The subsequent motion of the particle was assumed to be analogous to that of the primary nucleon.

12. The depth of the potential well for created π mesons was taken equal to 25 MeV according to [3]. Reflection and refraction was calculated in the same way as for nucleons.

13. The momentum balance was found after the calculation of each cascade and the momentum of the recoil nucleus was evaluated. This momentum was used to determine the kinetic energy of the nucleus and the direction of its recoil with respect to the direction of the incident beam.

14. The excitation energy of the residual nucleus was found from the energy balance, including the kinetic energy of the recoil nucleus.

15. The positions of collisions between cascade nucleons and the nucleons of the nucleus were recorded if the nucleon of the nucleus changed its momentum by more than 6 MeV.

The calculations according to the above scheme were performed on the M-20 computer.

A flow diagram of the computation is shown in the Fig. 1. At the start, operator 1 selects the point of entry into the nucleus and the energy and direction of the incident nucleon. Operator 2 then calculates the length of the nucleon path through nuclear matter and operator 3 determines the position of a collision. If the path length of the particle is greater than the radius of the nucleus, the particle is assumed to have escaped and control is transferred to operator 15. In the opposite case, we proceed to operator 4. The latter selects the collision partner from the nucleons present in the nucleus.

Operator 5 determines the type of the collision. If it is a nucleon–nucleon collision, we proceed to operator 6, whereas in the case of a pion–nucleon collision we proceed to operator 7. Operator 6 or 7 calculates the kinematics of the collision, while operator 9 checks that the Pauli principle is satisfied. If the collision can take place, we proceed to operator 10, otherwise control is transferred back to operator 2.

Operator 10 determines which of the particles have been absorbed and records the characteristics of the remaining particles for further processing. Operator 11 checks if there are enough memory locations left to record particle characteristics. It terminates the program when storage capacity is exceeded (this did not occur in practice), otherwise it stores the particle characteristics (position of the collision, kinetic energy, direction, type of particle). Operator 12 checks whether the storage area is empty and, when the area is full, calculates the characteristics of the particle and transfers control to operator 2. If there are no recorded particles, control is transferred to operator 13 which processes the results of one cascade (determines the kinetic energy and the excitation energy of the nucleus) and transfers control back to operator 1 if the number of cascades processed is less than given maximum number, otherwise it transfers control to operator 16.

Operator 14 calculates the escape of particles from the nucleus. If a particle leaves the nucleus, control is transferred to operator 15, if it is reflected, to operator 2. Operator 15 stores the characteristics of the emitted particles, while operator 16 processes the data from all of the computed cascades and constructs the angular and energy distributions of the emitted and knock-on particles and the distribution of knock-on nucleons throughout the volume of the nucleus.

The following quantities are determined from the calculation.

1. The average number of particles of each type emitted in a single cascade.

2. The angular and energy distributions of all emitted particles.

3. The average excitation energy of the residual nucleus.

4. The dependence of the excitation energy on the number of recorded collision points.

5. The dependence of the number of recorded collision points on the number of nucleons emitted by the nucleus.

6. The distribution of the collisions points inside the nuclear volume.

7. The excitation energy of the residual nucleus in cascades with a given number of emitted nucleons.

A description and analysis of the results obtained by the above method are given in [4-7].

LITERATURE CITED

1. N. Metropolis, R. Bivins, M. Storm, I. M. Miller, G. Friedlander, and A. Turkevich, Phys. Rev., 110:185 (1958).
2. R. Hofstadter, Phys. Rev., 28:214 (1956).
3. N. Metropolis, R. Bivins, M. Storm, I. M. Miller, G. Friedlander, and A. Turkevich, Phys. Rev., 110:204 (1958).
4. F. P. Denisov, Yu. A. Fatovskii, T. D. Kruglova, and R. A. Latypova, Preprint No. A-3 [in Russian], P. N. Lebedev Physics Institute (FIAN) (1962).
5. F. P. Denisov, R. A. Latypova, V. P. Milovanov, and P. A. Cherenkov, Preprint No. A-81 [in Russian], P. N. Lebedev Physics Institute (FIAN) (1964).
6. F. P. Denisov, R. A. Latypova, V. P. Milovanov, and P. A. Cherenkov, Yad. Fiz., Vol. 1, No. 2 (1965).
7. F. P. Denisov, R. A. Latypova, V. P. Milovanov, and P. A. Cherenkov, Yad. Fiz., Vol. 2, No. 5 (1965).

BEAM BEHAVIOR IN A SECTOR CYCLOTRON

A. T. Matachun, V. A. Gladyshev, L. N. Katsaurov, E. M. Moroz, and L. P. Nechaeva

Introduction

An electronic digital computer can be used to perform a complete analysis of particle motions in the magnetic field of a cyclic accelerator. These calculations can be used to solve the following problems:

a) harmonic analysis of the magnetic field;
b) determination of equilibrium orbits;
c) calculation of betatron-oscillation frequencies and the determination of the admissible betatron-oscillation amplitudes;
d) calculation of ion trajectories during ion drift in nonuniform magnetic fields.

Values of the magnetic field in cyclindrical coordinates r and θ were specified by tables. Let us consider as examples several calculations in the case of a 300-keV sector cyclotron [1-3].

1. Harmonic Analysis of the Magnetic Field

Harmonic-analysis data are used for the calculation of betatron-oscillation frequencies [4-6].

In addition, the higher harmonics can be used to detect departures from the necessary azimuthal symmetry and to evaluate their effect on orbital stability [6].

The expansion of the field in a Fourier series can be written as

$$H(r, \theta) = H_0(r) + \Sigma [a_n(r) \cos n\theta + b_n(r) \sin n\theta] = H_0(r) \{1 + \Sigma [A_n(r) \cos n\theta + B_n(r) \sin n\theta]\} = H_0(r) + \Sigma c_n(r) \sin(n\theta + \alpha_n). \quad (1)$$

The relations between the coefficients of these expansions are

$$A_n = \frac{a_n}{H_0}, \quad B_n = \frac{b_n}{H_0}, \quad c_n = \sqrt{a_n^2 + b_n^2}, \quad \tan\alpha_n = \frac{a_n}{b_n}. \quad (2)$$

Any of these formulas can be used, although the first is the more general one and the coefficients of the other two can be easily derived from those of the first.

The computer program for the harmonic analysis of the field calculates the Fourer coefficients from the given tabular values of the field strength. The values of the magnetic field for $0 \leq \theta \leq 2\pi$ are selected at a given value of the radius r and then the Fourier coefficients are calculated from the formulas

$$H_0(r) = a_0 = \frac{1}{2\pi} \int_0^{2\pi} H(r, \theta) d\theta, \quad a_n = \frac{1}{\pi} \int_0^{2\pi} H(r, \theta) \cos n\theta \, d\theta,$$

$$b_n = \frac{1}{\pi} \int_0^{2\pi} H(r, \theta) \sin n\theta \, d\theta, \qquad (3)$$

where $H(r, \theta)$ is the magnetic field strength for r = const.

A harmonic analysis program for the M-20 computer has been written at the P. N. Lebedev Physics Institute. The input to the program is in the form of a table giving the magnetic field at 2° intervals of θ for r = 1, 2, 3, ..., 32 cm. A library program for the evaluation of integrals by Simpson's rule is used for the calculation of Fourier coefficients. The output from the program is a printed list of expansion coefficients.

2. Determination of Equilibrium Orbits

The determination of an orbit reduces to the solution of the equations of motion of a charged particle in a given magnetic field under the condition that the orbit must be closed on itself at the end of one particle revolution. The equation of motion of a charged particle in a magnetic field in the median plane can be written as

$$r'' = r + 2\frac{r'^2}{r} - \frac{(r^2 + r'^2)^{3/2}}{r} K. \qquad (4)$$

Here, the prime denotes differentiation with respect to the angle θ, i.e.,

$$r' = \frac{dr}{d\theta}.$$

The quantity K is defined by

$$K = \frac{e}{\sqrt{2mc}} \frac{H(r, \theta)}{\sqrt{W}}, \qquad (5)$$

where e is the particle charge, m its mass, and c is the velocity of light.

If H is expressed in oersteds and W in electron volts, then in the case of a deuteron we have

$$K = 0.0048961 \frac{H(r, \theta)}{\sqrt{W}}. \qquad (6)$$

The closure of the solution can be expressed as

$$r(\theta) = r(\theta + 2\pi), \quad r'(\theta) = r'(\theta + 2\pi). \qquad (7)$$

After the equilibrium orbits have been found, the ion-revolution frequency is calculated from its definition, namely,

$$\omega = \frac{2\pi c}{\int dl}\left(1 - \frac{3}{4}\frac{W}{E_0}\right)\sqrt{\frac{2W}{E_0}}, \qquad (8)$$

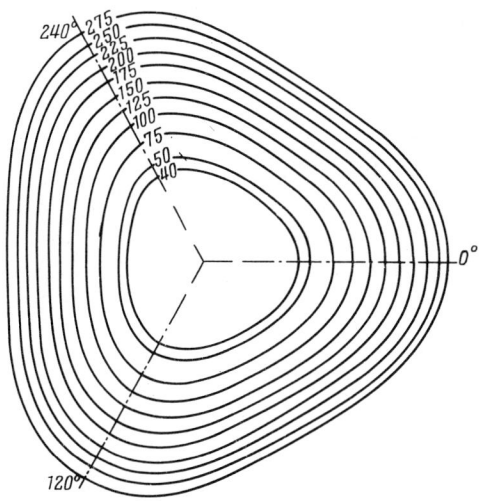

Fig. 1. Equilibrium orbits.

where $E_0 = 1.8755 \cdot 10^9$ eV and $c = 2.9979 \cdot 10^{10}$ cm/sec, $dl = \sqrt{(dr)^2 + (rd\theta)^2}$ and the integral $\int dl = \sum l_i$ is taken along the equilibrium orbit, l_i being an element of length. The average magnetic field along the orbit is calculated from

$$\overline{H} = \frac{\Sigma H_i l_i}{\Sigma l_i}, \qquad l_i = \sqrt{r_i^2 + r_i'^2}\,\Delta\theta_i. \tag{9}$$

The integration of Eq. (4) with the use of the closure condition (7) is a nonlinear boundary-value problem. Equation (4) was integrated numerically by the Runge-Kutta method with automatic step-length selection to keep the error below 10^{-5}. The boundary-value problem was solved by Newton's method.

The condition that the orbit is closed becomes

$$r(\theta_0) - r(2\pi + \theta_0) = f_1, \qquad r'(\theta_0) - r'(2\pi + \theta_0) = f_2, \tag{10}$$

where $r(2\pi + \theta)$ is a function of the two variables r_0 and r_0' specified at $\theta = 0$.

Let us use the abbreviations $r(\theta_0) = r_0$ and $r(2\pi + \theta_0) = r_k$. The system of equations (10) can then be rewritten as

$$r_0 - r_k(r_0, r_0') = f_1, \qquad r_0' - r_k'(r_0, r_0') = f_2. \tag{11}$$

Let us write down the system of equations for the corrections Δr and $\Delta r'$,

$$\frac{\partial f_1}{\partial r}\Delta r + \frac{\partial f_1}{\partial r'}\Delta r' = f_1, \qquad \frac{\partial f_2}{\partial r}\Delta r + \frac{\partial f_2}{\partial r'}\Delta r' = f_2, \tag{12}$$

where

$$\frac{\partial f_1}{\partial r} = 1 - \frac{\partial r_k}{\partial r} = 1 - \frac{r_k(r_0 + \Delta r, z_0') - r_k(r_0, r_0')}{\Delta r},$$

$$\frac{df_1}{dr'} = -\frac{\partial r_k}{\partial r'} = -\frac{r_k(r_0, r_0' + \Delta r') - r_k(r_0, r_0')}{\Delta r'},$$

$$\frac{\partial f_2}{\partial r} = -\frac{\partial r_k'}{\partial r} = -\frac{r_k'(r_0 + \Delta r, r_0') - r_k'(r_0, r_0')}{\Delta r},$$

$$\frac{\partial f_2}{\partial r'} = 1 - \frac{\partial r_k'}{\partial r'} = 1 - \frac{r_k'(r_0, r_0' + \Delta r') - r_k'(r_0, r_0')}{\Delta r'}. \tag{13}$$

System (12) which is linear in Δr and $\Delta r'$ is solved by Newton's method. This method exhibits good convergence only when the initial approximation is sufficiently close to the root. Therefore, it is desirable to specify the initial values of r and r' at the point $\theta = 0$ as close as possible to the true values.

Since the magnetic field $H(r, \theta)$ is given in tabular form, it is necessary to carry out an interpolation over θ and r during the integration of Eq. (4). The Lagrange interpolation formula for h = const is used for this purpose, namely,

$$y = \frac{[x-(a+h)][x-(a+2h)]}{2h^2} y_0 + \frac{(x-a)[x-(a+2h)]}{-h^2} y_1 + \frac{(x-a)[x-(a+h)]}{2h^2} y_2. \tag{14}$$

The program for the determination of an equilibrium orbit finds the equilibrium orbit for the energy specified, prints the details of the orbit, and then calculates the average magnetic field along it. Computation time is 2-3 min per orbit. As an example, Fig. 1 shows the equilibrium orbits in a sector cyclotron for deuterons with energies of 40, 50, 75, 100, 125, 150, 175, 200, 225, 250, and 275 keV.

3. Orbit Stability, Betatron Oscillations

The properties of a given orbit can be established by the solution of Eq. (4) with initial values that represent perturbations from an equilibrium orbit. Such calculations allow us to calculate the frequencies of betatron oscillations and to establish the limit on the oscillation amplitude such that when it is exceeded buildup of oscillations follows and the particle moves away from the equilibrium orbit.

Let us first of all consider the radial betatron oscillations. Let p_r be the radial component of the ion momentum for an equilibrium orbit, p_r^* the radial component of the ion momentum for the perturbed orbit, and let us denote the difference $p_r^* - p_r$ by Δp_r.

Since we have $r' = r \tan \xi$, where ξ is the angle between the trajectory and a circle, and

$$p_r = p \sin \xi = mv \sin \xi = mv \frac{\frac{r'}{r}}{\sqrt{1 + \left(\frac{r'}{r}\right)^2}}, \tag{15}$$

we find that

$$\Delta p_r = mv \left[\frac{\frac{r^{*'}}{r^*}}{\sqrt{1 + \left(\frac{r^{*'}}{r^*}\right)^2}} - \frac{\frac{r'}{r}}{\sqrt{1 + \left(\frac{r'}{r}\right)^2}} \right], \tag{16}$$

where r and r' refer to the equilibrium orbit. When ξ is small, so that we have $\sin \xi = \xi$, the quantity Δp_r is proportional to the angle between the equilibrium and perturbed orbits.

We now solve Eq. (4) with the given initial values of Δr_0 and Δp_r. When Δr_0 and Δp_0 are small, the perturbed trajectory is close to the equilibrium one and the particle performs periodic oscillations about the equilibrium orbit.

For each value of the azimuth we can construct a "phase diagram" in which Δr is plotted along the axis of abscissas and the value of Δp_r after each revolution (i.e., after the machine follows the trajectory through an angle 2π) is plotted along the axis of ordinates. The values of Δr and Δp_r for a given value of θ are plotted on a diagram. After each revolution, a point on the phase diagram shifts by an angle governed by the phase shift in the radial oscillations. The phase shift after N_r revolutions will be 2π and the point will trace a closed ellipse in the "phase diagram." The semiaxes of this ellipse represent the amplitude of radial betatron oscillations and the amplitude of the oscillations in Δp_r for the given azimuth and the given initial perturbation. The number of revolutions N_r which result in the closure of the ellipse is related to the betatron-oscillation frequency by the relation $N_r = (Q_r - 1)^{-1}$, where $Q_r = \omega_r/\omega$. Thus,

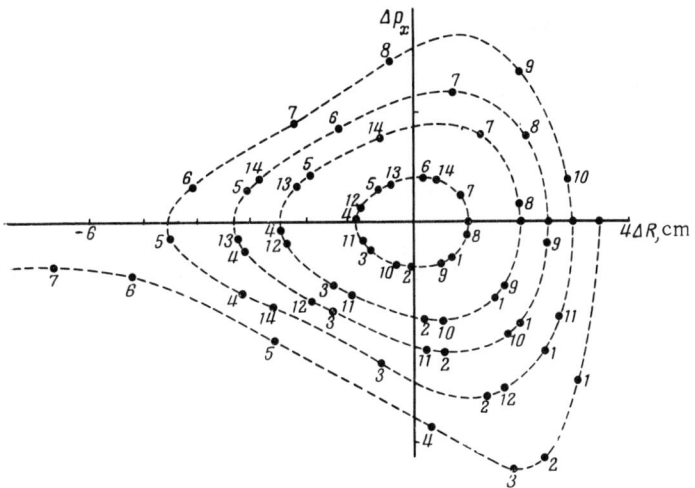

Fig. 2. Phase diagram for radial motion of deuterons with an energy of 50 keV; the numbers on the curves denote the revolution numbers.

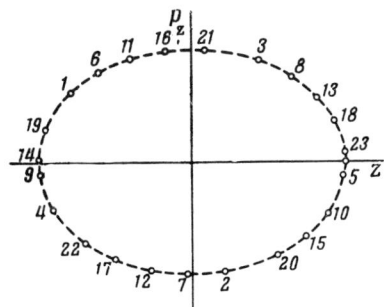

Fig. 3. The phase ellipse for vertical oscillations at an energy of 125 keV; numbers on the curve denote revolution numbers.

having calculated N_r revolutions, we obtain the frequency of radial betatron oscillations.

As the initial deviations from an equilibrium orbit get bigger, the shape of the phase ellipse becomes more and more distorted, i.e., the oscillations become increasingly nonlinear, until having reached a critical amplitude, the oscillations become unstable and the deviation of the orbit from an equilibrium one increases without limit. There is also a nonlinearity in the dependence of the betatron-oscillation frequency on the amplitude. Calculations of this type allow us to estimate the extent of the domain of stable oscillation amplitudes and, in the case of traversal of a resonance region, to determine the maximum number of revolutions needed to traverse the dangerous region.

Phase diagrams calculated for the 50-keV sector cyclotron of the Institute of Physics are shown in Fig. 2.

4. Analysis of Vertical Motion

The vertical motion can be analyzed by means of phase diagrams in the same way as in the radial betatron oscillations.

Let us specify the initial deviation from the median plane on the equilibrium orbit and let us solve Eq. (4) simultaneously with the equation for the axial motion

$$z'' = z' \frac{r'r'' + rr'}{r^2 + r'^2} - z \frac{K}{H} (r^2 + r'^2)^{1/2} \left(\frac{r'}{r} \cdot \frac{\partial H}{\partial \theta} - r \frac{\partial H}{\partial r} \right), \qquad (17)$$

where $z' = dz/d\theta$.

In constructing the ellipse after each revolution (i.e., after the equation has been integrated over an angle of 2π) along the axis of abscissas we plot the value of z, the deviation from the median plane at the given azimuth, and along the axis of ordinates we plot p_z, the magnitude of the axial momentum (the axial momentum is zero on an equilibrium orbit lying in the median plane). p_z is proportional to the sine of the angle φ between the perturbed trajectory and the median plane. The value of $\sin \varphi$ can be calculated from

$$\sin \varphi = \frac{z'}{\sqrt{r^2 + r'^2 + z'^2}}. \tag{18}$$

In deriving formula (17), we have assumed that the radial and azimuthal components of the magnetic field strength depend linearly on z, so that nonlinearities do not manifest themselves in the construction of the ellipses. The phase ellipse for axial oscillations at an energy of 125 keV is shown in Fig. 3. The Runge-Kutta method was used to integrate the system of equations (4) and (17). The partial derivatives $\partial H/\partial \theta$ and $\partial H/\partial r$ were calculated numerically from the given tabular values of $H(r, \theta)$.

The number of revolutions N_z after which the phase ellipse becomes closed is related to the number of axial betatron oscillations Q_z by the relation

$$N_z = (1 - Q_z)^{-1}. \tag{19}$$

With the help of a digital computer it is possible to obtain the betatron-oscillation frequencies by a somewhat different method. If we assume that the betatron oscillations are described by linearized equations, the radial and axial oscillations being separable, then the solution of Hill's equations for the radial deviations from the equilibrium orbit can be represented as

$$x = 2a \, |f(\theta)| \cos [Q_r \theta + \psi(\theta) + \beta]. \tag{20}$$

This is called the normal norm of the Floquet solution. Here, a and β are real constants and $\psi(\theta) = \arg f(\theta)$ is a periodic function.

An analogous expression is obtained for the axial oscillations. If we consider oscillations at a given azimuth, i.e., for a fixed value of θ, then the consecutive deviations from an equilibrium orbit x_1, x_2, x_3 will be described by a cosine law, while the value of $f(\theta)$ for a given value of a which depends on the initial conditions will characterize the maximum deviation at the given azimuth. Having calculated three orbits on the machine for a given initial deviation, we will have at each azimuth three consecutive values of the deviation, x_1, x_2, and x_3, which must lie on a cosine curve, i.e., they must satisfy the equations

$$\begin{aligned} x_1 &= A \cos [Q_x \theta + \beta], \\ x_2 &= A \cos [Q_x (\theta + 2\pi) + \beta], \\ x_3 &= A \cos [Q_x (\theta + 4\pi) + \beta]. \end{aligned} \tag{21}$$

The constant A represents the oscillation amplitude at the given azimuth. If we calculate the amplitude at all angles, we will obtain the envelope of the betatron oscillations corresponding to the given initial perturbation. We can easily derive formulas from Eqs. (21) for the calculation of the number of radial betatron oscillations suffered by a particle during one revolution, namely, $Q_x = \omega_x/\omega$, where ω_x is the frequency of radial betatron oscillations and ω the angular frequency of the particle,

$$Q_x = P_x + \alpha_x, \tag{22}$$

$$2\pi \alpha_x = \pm \arccos \frac{x_3 + x_1}{x_2}. \tag{23}$$

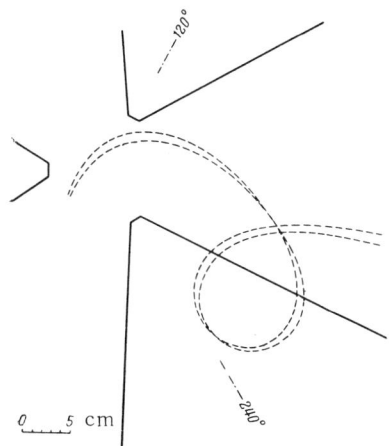

Fig. 4. Two adjacent trajectories showing the changes in the dimensions of the beam in the median plane during drift.

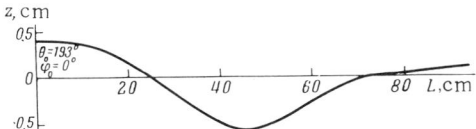

Fig. 5. The magnitude of the deviation from the median plane calculated along the trajectory; θ_0 is the azimuth of the injection point, φ_0 the initial value of the angle between the injected beam and a radius.

The magnitude of P_x and the sign of α_x is given by the following rule from the number m_x of zeros of the function $x(\theta)$ in the interval $0 < \theta < 6\pi$.

$$\begin{aligned} \text{If} \quad & m_x < 3, \text{ then } P_x = 0, \; \frac{\alpha_x}{|\alpha_x|} = +1, \\ \text{if} \quad & 3 \leqslant m_x \leqslant 6, \text{then } P_x = 1, \; \frac{\alpha_x}{|\alpha_x|} = -1, \quad (24) \\ \text{if} \quad & m_x > 6, \text{ then } P_x = 1, \; \frac{\alpha_x}{|\alpha_x|} = 1. \end{aligned}$$

This method is suitable for the calculation of the frequencies of radial and axial oscillations. By comparison with the method described in sections 3 and 4, this method has the advantage of speed. Only three revolutions have to be calculated for the determination of Q_r and Q_z. The values of Q_r and Q_z obtained in this way are not averaged over azimuth and depend (although weakly) on the selected initial conditions.

5. The Calculation of Trajectories for Ion Drift in an Inhomogeneous Magnetic Field

The solution of Eqs. (4) and (17) allows us to study ion drift in inhomogeneous magnetic fields, so that we can determine the character of focusing in the vertical and median planes and find the positions of the focal points. There is no difference in principle between this and preceding problems.

If the trajectory forms loops because of drift, the derivative goes through infinity. In this case, the coefficient K appearing in Eqs. (4) and (17) contains an additional factor $\chi = \pm 1$, i.e., we now have

$$K = 0.0048961 \frac{H(r, \theta)}{\sqrt{W}} \chi,$$

where χ changes sign each time r' goes through infinity.

Figure 4 shows the calculated trajectory of deuterons injected into a sector cyclotron in the median plane.

The vertical deviation from the median plane due to drift in an inhomogeneous magnetic field is shown in Fig. 5.

LITERATURE CITED

1. V. A. Gladyshev, L. N. Katsaurov, A. N. Kuznetsov, L. P. Martynova, and E. M. Moroz, Proceedings of the International Conference on Accelerators, Dubna [in Russian], Atomizdat, Moscow (1963), p. 658.

2. V. A. Gladyshev, L. N. Katsaurov, A. N. Kuznetsov, L. P. Martynova, and E. M. Moroz, Atom. Énerg., 18:213 (1965).
3. V. A. Gladyshev, L. N. Katsaurov, A. N. Kuznetsov, E. M. Moroz, and L. P. Nechaeva, Atom. Énerg., 11:443 (1965).
4. G. Parzen, Sector-Focused Cyclotrons, Proceedings of an Informal Conference, Sea Island, Georgia (1959), p. 40.
5. R. G. Bassel and R. C. Bender, Nucl. Instr. Meth., 6:34 (1960).
6. H. L. Hagedoorn and N. F. Verstor, Nucl. Inst. Meth., 18/19:201 (1962).

VASSAL
(Automatic Memory Allocation System)

U. G. Martin

Introduction

The language Vassal is a system for automatic memory allocation in the machine M-20 when a teletype input devise is available. The principles utilized in this language are similar to those in which certain earlier automatic programming languages (called "Assemblers") were based on the USA and may be easily adapted to any computer. The language Vassal is substantially different from automatic memory allocation systems currently used in the USSR [1, 2]. Since the need for hand programming will not disappear in the near future, simple automation systems of the type of Vassal will find useful application.

Vassal is intended for the programming of small and medium volume problems. Furthermore, the present form of the translator can program problems in length up to $243_{(10)}$ punch cards or up to $23328_{(10)}$ teletype symbols. This makes a machine program with maximum length about $1300_{(8)}$ instructions, not including memory locations for data.

A program in Vassal has up to $617_{(8)}$ different labels, where each one may used in the program an arbitrary number of times. The translator for this language is faster, since the M-20 backup store is not used.

Vassal is most useful in those cases where ALGOL is either impossible or difficult to use, for example: in the writing of procedures, codes, translators, certain standard subroutines, and the solution of data-processing problems. The volume of the majority of such problems does not exceed that stated above, or can be divided into parts in such a way that each one does not exceed the limit.

It is easier to write and understand a program in Vassal than one in machine language. This is explained by the fact that the labels form short words and abbreviations, which may be connected with the instructions and constants to which they are attached in the program. A language of the type of Vassal is very convenient in program debugging. In hand programming as well as in symbolic-language programming errors are frequently due to inattention on the part of the operator. With a language of the type of Vassal such errors are most often detected automatically during translation by a diagnostics block and are corrected by the translator. A substantial advantage of programs written in Vassal type languages is also that they may be automatically transferred to any location in core.

1. Input Language for the Vassal Translator

Programming in Vassal is very similar to programming in machine language. The basic difference consists in the following: any instruction or constant, indeed, practically any register in core may be assigned a label which is then used for access to the given register.

A label is a sequence of not more than five arbitrarily disposed Latin letters and numbers, beginning with a letter. This definition differs from that adopted in ALGOL only in that the length is limited to five symbols. A label in Vassal combines the properties of label and identifier in ALGOL, so that sometimes a label is used in Vassal to transfer control or to indicate the variable contained in a memory register.

A program written in Vassal is called symbolic, and the machine language program obtained from the translator is called absolute. Each instruction in a symbolic program is called symbolic.

Aside from the ordinary instructions of the M-20, Vassal has two literal pseudoinstructions. Each symbolic program must begin with a symbolic instruction that contains the pseudoinstruction "SP" (start of program) and an octal number in the second address field. This instruction is the "start instruction," and the number in the second address field indicates to the translator at which memory register the instruction of the absolute program starts. After translation the start instruction "vanishes," i.e., does not appear in the absolute program.

A second pseudoinstruction is "rm" ("reserve array" or simply, the instruction "reserve"). By this instruction the translator locates a successive group of registers for the reception of input data or for working registers. If the required number of registers is equal to octal n, in the second address of the instruction "reserve" the parameter n − 1 is written (for example, on the instruction "reserve" with zero in the second address the translator locates one register). The instruction "reserve" may be tagged with an arbitrary label, which is taken analogously to the array identifier in ALGOL, and may be considered as a coarse analogy to the array descriptor.

Each "reserve" instruction appears in the absolute program as a zero word, independently of the parameter contained in the second address. However, memory is allocated as if the "reserve" instruction occupied n registers. If all "reserve" commands are written in order at the end of the symbolic program, then the absolute program is printed in a form not requiring further processing. It is therefore recommended to follow this method.

In principle the "reserve" commands can be located arbitrarily in the symbolic program. Unfortunately, however, the address code KA cannot be generated on the M-20 punch. Therefore, if the programmer has neglected the above recommendation, and has placed "reserve" instructions freely in the symbolic program, those zero words which substitute each "reserve" instruction must be corrected and changed in absolute program to KA, except for the case where only zeros follow the command "reserve." The translator can be composed in such a way that the number of zeros formed is equal to the length of the array. But in many cases this would lead to printing and punching a large number of unnecessary zeros. This would complicate the exploitation of the translator.

Each instruction of the symbolic program consists of five parts (or "fields"). This is the field of labels or the left column (up to five teletype symbols), the field KOP (up to three teletype symbols) and the address field (up to eight teletype symbols in each address, consisting of not more than five symbols for the label or absolute address, and not more than three symbols for the shift sign, which will be described below). The address fields of the symbolic instruction may be used in four different ways.

It is possible to write in the address:

1) up to four octal numbers, which will be read as an absolute address or a part of an octal constant, where the number will be put into the absolute program unchanged;

2) a label, which must be found once and only once in the left column of the symbolic program and will be substituted in translation by the number representing the address of this label with respect to the prescribed start of the symbolic program;

3) the "plus" or "minus" sign, followed by one or two octal numbers (the so-called "address formation with shift" — afws), where this entire group of symbols will be substituted by a number representing the future absolute address of the register where it will be written, plus or minus the number of registers indicated after the "plus" or "minus" sign.

4) a label followed by afws, which will be substituted by the number representing the future address of this label, plus or minus the shift indicated in afws.

It is easy to prepare special blanks for programming in Vassal. It is possible, although this is not at all convenient, to use the ordinary M-20 programming blanks. For this it is necessary to write very small, and to use as the label field the left column of the blank, usually used for the addresses. The labels of maximum permissible length are placed in the address fields, although it is not often required to use labels of maximum length. Labels of one or two symbols are perfectly adequate for the majority of cases.

Let us give an example of a symbolic program written in Vassal, with subsequent translation to machine language. The program which we have taken as the example was written by V. A. Fomina in machine language for the calculation of three groups of integrals according to the standard program written by F. I. Strizhevskaya.

The problem is formulated in the following way: compute

$$\Delta_i = \sqrt{(\Delta C_i)^2 + (\Delta S_i)^2} \quad \text{for } i = 1, 2, 3, 4,$$

where

$$\Delta C_i = \int_0^\pi \{[\sin(N_i \sin\psi)]\sin\psi\}\,d\psi, \qquad \Delta S_i = \int_0^\pi \{[\cos(N_i \sin\psi)]\sin\psi\}\,d\psi,$$

$$\Delta C_i = \int_0^{\pi/2} \{[\sin(N_i \sin\psi)]\sin 2\psi\}\,d\psi, \qquad \Delta S_i = \int_0^{\pi/2} \{[\cos(N_i \sin\psi)]\sin 2\psi\}\,d\psi,$$

$$\Delta C_i = \int_0^{\pi/4} \{[\sin(N_i \sin\psi)]\sin 4\psi\}\,d\psi, \qquad \Delta S_i = \int_0^{\pi/4} \{[\cos(N_i \sin\psi)]\sin 4\psi\}\,d\psi$$

and where $N_1 = \pi$, $N_2 = 2\pi$, $N_3 = 4\pi$, and $N_4 = 6\pi$.

Let us consider first the symbolic program.

The lines given in the example to indicate the contents and the points may be omitted in writting the program.

.		SP	.	0010	.	v2	.	.	.	.
.		16	+1	7501	7610	.		52	.	.	.
.		52	w0	0042	r2	.	6	05	r25	r25	r13
.	.		7761	.	r3	ds	12 6	05	r26	r26	r14
v4		16	+1	7501	7610	.		01	r13	r14	r15
.		75	w0	0152	x0	.	1	44	r15	.	r16
.		75	*cabl*	v1	*ea*	.	1	12	0003	v2+2	0001
.		75	*kobi*	v2	*pi*	.		16	+1	7501	7610
.		75	n	r22	n1	.		52	r16	0027	r21
.		75	k	koba	k1	v5		.	.	.	.
.		16	.	v2+1	.	.		05	7762	r3	r3
v1		05	r3	r22	r5	.		04	*pi*	7762	*pi*
.		16	+1	7501	7610	.		16	+1	v4	−3
.		75	r5	0005	r6	.		05	7762	r3	r3
.		16	v3−1	7501	7610	.		04	*pi*	7762	*pi*
.		75	ч22	0005	r7	.		56	+1	v4	v5
v3		52	.	.	.	.		37	.	.	.
.	4	00	*pit*	.	r4	n		0010	.	.	
.		05	r4	r7	r10	n1		0007	.	.	
.		16	+1	7501	7610	k		.	.	.	
.		75	r10	0071	r11	k1		.	.	.	
.	1	05	r11	r6	r23	koba	1	01	4000	.	.
.	1	05	r12	r6	r24	.		16	+1	7501	7610
.	1	12	0003	**v3**+1	0001	.		52	r3	0027	r27
.						.		37	.	.	.

Constants

w0	.	.	.	.
x0	.	.	.	.
eobl	.	.	.	.
ea	.	.	.	.
kobi	.	.	.	.
pi	.	.	.	.
pit	.	.	.	.
pi2	.	.	.	.
pi4	.	.	.	.
pi6	.	.	.	.
r1	.	.	.	.
r2	.	.	.	.
.	rm	.	0073	.

Working registers

r3	.	.	.	.
r4	.	.	.	.
r5	.	.	.	.
r6	.	.	.	.
r7	.	.	.	.
r10	.	.	.	.
r11	.	.	.	.
r12	.	.	.	.
r13	.	.	.	.
r14	.	.	.	.
r15	.	.	.	.
r16	.	.	.	.
r17	.	.	.	.
r20	.	.	.	.
r21	.	.	.	.
.	.	.	.	.
.	rm	.	0057	.

Working registers

r22	.	.	.	
r23	.	.	.	
.	.	.	.	
.	.	.	.	
r24	.	.	.	
.	.	.	.	
.	.	.	.	
r27	.	.	.	
.	.	.	.	
.	.	.	.	

.	.	.	.	
.	.	.	.	
.	.	.	.	
.	.	.	.	
.	.	.	.	
.	.	.	.	
r25	.	.	.	
.	.	.	.	
.	.	.	.	
r26	.	.	.	

In machine language, the program appears as

0010	016	0011	7501	7	610	0040	052	0000	0000	0	000
	052	0070	0042	0	103		005	0321	0321	0	210
	000	7761	0000	0	200		005	0325	0325	0	211
	016	0014	7501	7	610		001	0210	0211	0	212
	075	0070	0152	0	071		144	0212	0000	0	213
	075	0072	0022	0	073		112	0003	0041	0	001
	075	0074	0037	0	075		016	0047	7501	7	610
	075	0060	0300	0	061		052	0213	0027	0	216
	075	0062	0064	0	063		000	0000	0000	0	000
	016	0000	0040	0	000		005	7762	0200	0	200
	005	0200	0300	0	202		004	0075	7762	0	075
	016	0024	7501	7	610		016	0054	0013	0	050
0024	075	0202	0005	0	203	0054	005	7762	0200	0	200
	016	0026	7501	7	610		004	0075	7762	0	075
	075	0300	0005	0	204		056	0057	0013	0	050
	052	0000	0000	0	000		037	0000	0000	0	000
	400	0076	0000	0	201		000	0010	0000	0	000
	005	0201	0204	0	205		000	0007	0000	0	000
	016	0033	7501	7	610		000	0000	0000	0	000
	075	0205	0071	0	206		000	0000	0000	0	000
	105	0206	0203	0	301		101	4000	0000	0	000
	105	0207	0203	0	305		016	0066	7501	7	610
	112	0003	0030	0	001		052	0200	0027	0	310
	000	0000	0000	0	000		037	0000	0000	0	000

2. Some Remarks on Programming in Vassal

Except for decimal numbers, which may be encountered in the labels, the symbolic program is written in the octal system. This means that decimal numbers are only used as input data or are pretranslated into the octal system.

In Vassal programming it is recommended to have four blanks: for writing the symbolic program, for the constants, for working registers, and for "reserve" commands. After a blank has been completed for any of these four groups, a fresh blank is taken for the same group. The current blank for the symbolic instructions must always be in view, and the other blanks in the instruction part of the program and the other three groups are uncovered to the extent that they contain information required for programming. Most often only the left column of the other blanks is required.

The lines of the symbolic program have labels as required, not necessarily in the order of writing the program.

Before giving the completed program to be punched all the blanks in the last three groups (as listed above) are numbered in sequence following the last number on the instruction part blanks. Where each of the four groups ends the remaining unused lines on the last blank should be crossed off by a single large "Z," indicating that they should be omitted in punching. If the program is written as described, the programmer need not concern himself with the final length of each group, and the translator allocates memory without gaps, so that the absolute program occupies the minimum number of successive registers.

The simple alteration of the pseudoinstruction "SP" at the start of the program permits the program to occupy any place in core in retranslation. This permits several programs to be combined into one, even when the programs have been independently written.

If an instruction or group of instructions has been omitted inadvertently in the writing of the program, in place of the usual "patch" the programmer can simply insert the missing instructions at the required place by repunching several cards on the teletype* and retranslating the program. Similarly, after the program has been debugged, the instructions which were included in the program only for debugging may be removed and the program retranslated. Program translation in Vassal is very fast, so that repeated translation of the program is perfectly economical, when required.

However, the following should be borne in mind. When an instruction is inserted or removed from the program it is necessary to check that those instructions close to the point of modification using address formation with shift (afws) are changed appropriately, if required by the sense of the algorithm.

3. Translator Operation

After the symbolic program has been punched on the teletype, a special card is added at the end of the program with the legend "Vassal — end of symbolic program." This card contains two lines, for carriage return and zero check sum.

The translator consists of 58 punched cards and check sum and may be stored on magnetic tape. In this case a card is placed in the reader for Vassal input from tape with one or several symbolic programs, each of which is given its end-of-program card. When the input button is pressed (it is not necessary to first erase core) and the translator is called in from tape, the first program will be translated. The next symbolic programs after the first stop of the translator are translated by pushing the button "Start" (the Vassal translator is not called in again from tape). The translator may also be brought into the machine by the reader; in this case it simply substitutes its calling card in the above rules.

The program is translated in three stages. During the first stage a "dictionary" of different labels is formed. This dictionary is then ordered by label, considering it as a number. The second stage is fictitious translation with binary search in the dictionary to detect various types of errors. Depending on the results of the second stage, various actions are carried out in the third stage.

Since the symbolic program does not have a check sum, the translator always computes it. If there is a suspicion that translation was incorrect due to incorrect reading of the program, the program is retranslated and the two check sums compared. At the start of operation the translator prints out the check sum indicator 010101010101010 and then the check sum.

If no errors are found during the fictitious translation, the indicator "OK" is printed, signifying "everything in order" (like all alphabetic indicators, it is composed of numbers [sic!]). Then the absolute program is printed in groups of 12_{10} instructions, where each of the groups has a line with indication of the address of the first instruction in the group. After each group of instructions has been printed it is punched. At the end the check sum of the program is punched. It will take into account the missing KA (corresponding to the pseudoinstruction "SP" at the start of the symbolic program), which the programmer must add at the start of the absolute program. In the same way, in the presence of a "reserve" command in the middle of the program, the impossibility of punching KA by the machine gives rise to inconveniences for the programmer.

If the translator has not found more than eight errors, the absolute program is printed and punched as described above, taking into account that for a small number of errors the pro-

---

* Apparently the Soviet term "teletype" does not correspond exactly to the well-known "Teletype" machines using punched paper tape — Translator.

grammer can easily correct them himself in the absolute program, rather than retranslate.

If more than eight errors have been found, the program is printed with the subsequent indicator "NG" ("no good"). In this case the absolute program is not punched, and it must be retranslated after error correction.

If the number of printed errors reaches 30_{10}, the indicator "Oh you!" is printed, and the translator stops in the state "ready to translate next program." In this case the absolute program is neither printed nor punched.

In all of these cases the translator prints each error in the form of a code number indicating the type of error and a line of information which (if present) gives additional information.

The translator may detect the following six types of errors:

Code number	Information line
1. 020202020202020. Dictionary size of 617_8 labels exceeded (the program may contain in the address parts many more labels than this, since the dictionary volume is defined by the number of distinct labels).	Absent
2. 030303030303030. In the address field beginning with an octal number a teletype symbol has been found which is not an octal number.	The absolute address of the erroneous instruction is given in the second address.
3. 040404040404040. After the "plus" or "minus" sign a teletype symbol has been found within the address which is not an octal number.	
4. 050505050505050. In one of the addresses of the instruction whose address is given in the information line, a label is used which does not appear anywhere in the label field.	
5. 060606060606060. At the address indicated in the information line a label is used which is given more than once in the label field.	The first three and last digits of the information line compose an absolute address of a doubtful label. The digits from the fourth to the eighth, inclusive, compose a label in six-bit symbols, where the last five bits of each symbol correspond to the positions of the international teletype code, and the first bit indicates the alphabetic or the numeric register (0 or 1, respectively).

6. 070707070707070.

At least one of the five fields of the symbolic instruction exceeds the permissible length or is not separated from the adjacent one by one (or several) spaces.

As in the second and fourth cases.

If an error is found in the address, in translation the address will be taken equal to zero.

One possible cause of a large number of errors is the absence of the end of symbolic program card. In this case the machine stops at KRA – 1177 and for translation of the next program control must be transferred to address 0012.

Other frequent causes of error are the following: incorrect orthography of labels, absence of label, absence of label with decimal numbers used in constants.

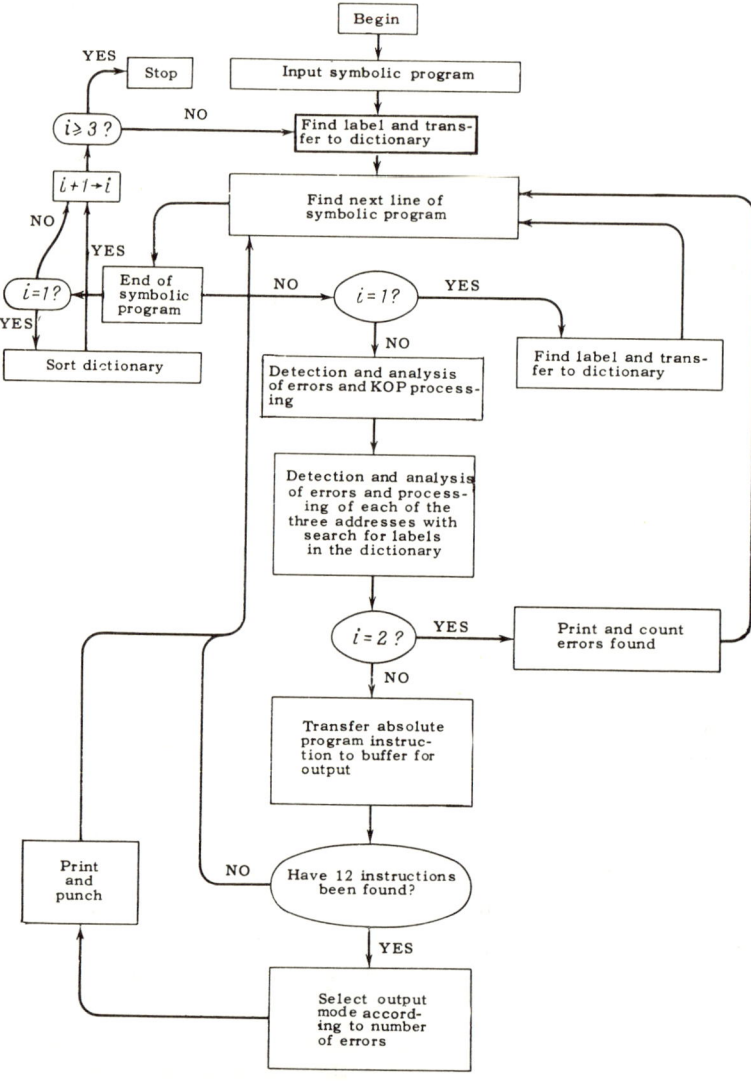

Fig. 1. Flowchart of the Vassal Translator.

4. Rules for Punching Vassal Programs

Each line of the symbolic program must contain not more than 36_{10} teletype symbols, exclusive of service codes.

Each of the five fields of the symbolic program must be separated from the adjacent ones by at least one space. Those fields of the symbolic program which the programmer has left empty must contain the teletype symbol " · " (point).

Each symbolic instruction is preferably punched with the following arrangement: five symbols for the label field, space, three symbols for the KOP field, space, and eight symbols for each of the three address fields and the spaces between them. Slight deviations from this arrangement will not be taken as errors, since the number of spaces between fields has no significance.

If during punching an incorrect key has been pressed, the corresponding line may be caused to be rejected by the translator by punching at any point of the line the multiplication sign "×" and the carriage return.

The figure gives a condensed flowchart of the Vassal translator. The program may be studied in greater detail if it is printed from the card deck.

LITERATURE CITED

1. V. V. Martynyuk, Library of Standard Programs [in Russian], edited by M. R. Shura-Bura, Computation Center of the Academy of Sciences of the USSR, Moscow, p. 198 (1961).
2. E. A. Zhogolev and V. I. Kurilov, Standard Program Assembler [in Russian], Moscow State University, Moscow (1964).

DETERMINATION OF REACTION CROSS SECTIONS FROM COUNTER-TELESCOPE DATA

L. I. Slovokhotov

Introduction

Measurements on a given reaction provide the experimenter with data in the form of an individual number, or a set of numbers, which we shall refer to as the output. The output is the number of counts recorded by counters, or the number of tracks in a chamber or on a photographic plate. The output is determined, on the one hand, by the reaction cross section, i.e., the probability of a particular process due to the interaction between an incident particle and a target nucleus or nucleon and, on the other, by the specific experimental conditions such as the size and efficiency of counters, the size of the target, the intensity of the incident particles, and so on. In order to deduce the cross section from the output one must first eliminate all the particular features of the experiment, i.e., introduce corrections for possible output losses due to secondary interactions of the reaction products with the target and the counters, taking into account the counter efficiency, and finally reduce the output to some standard conditions, for example, express it per unit solid angle, per target nucleus, and so on. In general, this is a relatively complicated problem. Thus, owing to ionization energy losses by the recorded charged particles, the effective volume of the target is not equal to its geometric volume, but is a complicated function of the experimental geometry and the incident-particle energy. A similar situation is encountered in the case of the solid angle within which the particles are recorded. The probability that a secondary particle will pass through different points in the detector is not a constant and, owing to the kinematic relation between the energy of the escaping particle and the angle of escape, it is a function of the incident-particle energy, the coordinates of the point of interaction in the target, and the energy discrimination in the counter telescope. The situation becomes exceedingly complicated when it is necessary to take into account multiple scattering of the incident particles during their passage through the target and counters. It is thus clear that an analytic approach to the solution of this problem is unlikely to be successful.

It is, however, possible to use numerical methods of calculation. Although the computational process is then a little more difficult to visualize, it is, nevertheless, possible to obtain a solution and to estimate the accuracy, although the problem has to be simplified somewhat since any attempt to obtain a more or less complete solution leads to an enormous rise in the number of necessary arithmetic operations. For example, when experimental data on angular distributions in the Compton effect were analyzed in 1960, using an approximate method to allow for multiple scattering [1], it took about four months of continuous calculations by four physicists using semiautomatic devices to analyze six angular-distribution points and the data of some control experiments. The analysis had to be carried out at least twice, first before the actual experiment, so that the optimum experimental parameters could be determined, and then again after the measurements.

This method is very advantageous if it is used in conjunction with a fast electronic computer. Its main advantages are as follows:

1. There is no need to introduce additional limitations on the number of arithmetic operations, and this substantially improves the results of the calculations.

2. By using a basic computational method, and varying only individual sections of the program connected, for example, with the kinematics of the process or the geometry of the experiment, it is possible to increase the number of problems and the number of possible formulations of the experiment. This means that optimal planning of the experiment becomes possible.

3. By varying different parameters it is possible to follow their effect on the final result, and to obtain extensive spatial and energy distributions of the recorded particles in individual counters or in individual collimators. In this way, both quality and quantity are improved and the numerical method, which has been so laborious in the past, acquires all the advantages of the analytic method, including the ease of vizualization.

4. It is possible to estimate the accuracy of the calculations and the resolution functions.

5. The use of electronic computers frees the experimenter from extremely time-consuming and tedious numerical work.

It is, of course, true that the use of computers will not result in the solution of all problems, and new difficulties arise which are specific to computer calculations.

In Chapter I we shall consider the formulation of the problem and the algorithm for calculating the cross section from the measured output for reactions occurring in a target exposed to a bremsstrahlung beam from a synchrotron. The secondary particles recorded by the counter telescope will be stable charged particles (protons). The computational scheme will, however, be valid for a broader range of problems in which the cross section has to be determined from some particular charged-particle output. If, however, the particles are unstable it is necessary additionally to investigate the probability of recording of these particles with allowance for the lifetime, the decay scheme, and the kinematics. We shall suppose that the detector is a counter telescope although, as will be seen below, the scheme is quite general and can be extended to certain other detection systems.

In Chapter II we shall consider, as an example, the analysis of experimental data on the Compton effect and the photoproduction of neutral pions on hydrogen. Here we shall use the particular characteristics of the hydrogen target, the synchrotron, the telescope, etc.

Finally, in Chapter III we shall give a block diagram and some individual elements of the program for the algorithm discussed in Chapter I, and apply it to the problems formulated in Chapter II. Moreover, we shall analyze some of the problems associated with the particular computational method which we have employed.

It is important to note that it will be assumed throughout that the telescope records particles produced only in a particular reaction, i.e., we shall not discuss the separation of a given reaction from other reactions which result in the same particle or particles at a given point.

CHAPTER I

Relation Between the Output and the Rear Section

§1. Formulation of Experiment. Definitions

The Counter Telescope. Counter telescopes consisting of proportional, scintillation or other counters are used to define the direction of motion of the particles. In addition, the telescope defines the energy interval of the recorded particles and, in general, the type of particle, i.e., it carries out mass separation.

For our particular problem, an important parameter of the telescope is its ability to record charged particles in a given energy interval, i.e., a given interval of ranges, in which the minimum energy is determined by the amount of material traversed by the particles up to the last coincidence counter. The energy expended by the particles in the last counter as a result of the detection process is neglected. This is a variable quantity since it depends on the sensitivity threshold of the corresponding electronic circuits. However, if it is an appreciable fraction of the minimum energy recorded by the telescope, it can be taken into account and included in each particular case (for example, through the "threshold" range in the last counter) in the minimum energy of the telescope. Moreover, it is assumed that the detection probability for particles within the chosen energy interval is constant, i.e., it is independent of both the particle energy and the point of entry into the counter. This assumption is not particularly fundamental because it can be removed at the expense of some complication. It follows that the necessary condition for particle detection is that the particles must fall into the energy (or range) interval of the telescope, and into the geometric volume of the counters.

The material which governs the energy threshold of the telescope is usually distributed nonuniformly in space, and is localized in certain special places, for example, in filters (absorbers), telescope or target walls, counters, and the various components of the telescope. It is therefore assumed in the computational scheme for the telescope that all the material is localized in a few scatterers whose mutual disposition, size, and thickness represent the distribution of matter and the geometry in the real telescope. The geometric thickness of the scatterers can be neglected in comparison with the distance between them; only their thickness, i.e., the amount of matter in them, is taken into account. Moreover, there are diaphragms representing the collimators or other constructional elements of the telescope which shape the charged-particle beam. The general computational scheme for the telescope is illustrated in Fig. 1. It is assumed that all the scatterers and diaphragms are coaxial.

The Spectrum and Flux of Bremsstrahlung. Target. In this chapter only some very general requirements will be imposed on the particle beam from the accelerator. The only necessary condition will be that there is a known analytic relation between the energy and the spatial distribution of the particles in the beam, and that the beam surface (if it is bounded) can be described by some analytic formula. As far as the target is concerned, all that is required is that it must be homogeneous and of known shape.

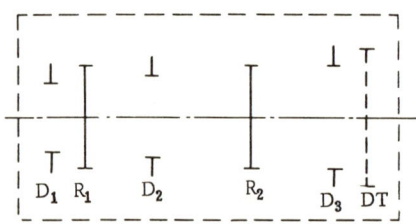

Fig. 1. General scheme for telescope calculations. R_i) Scatterers; D_i) diaphragms; DT) detector.

§2. Calculation of the Cross Section from the Output without Correction for Multiple Scattering

The geometry of the experiment is defined in Fig. 2. Consider a coordinate system x, y, z with the

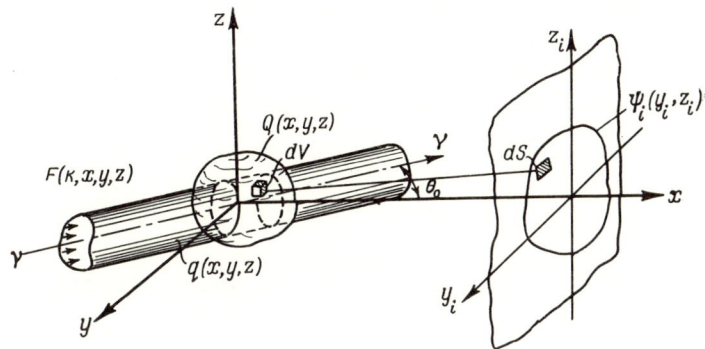

Fig. 2. Geometry of the experiment.

origin at some point in the target. The x axis will be taken to lie along the telescope axis, i.e., at right angles to the plane of the detector (and to the planes of the diaphragms).

Let $Q(x, y, z) = 0$ be the equation describing the surface of the target, $q(x, y, z) = 0$ the surface of the beam, and $\varphi_i(y_i z_i) = 0$ the contour of the i-th diaphragm or detector. By solving simultaneously the first two equations we find the equations for the lines of intersection between the surface of the target and the surface of the beam, and hence define the "active" volume of the target, i.e., the volume in which the primary particles interact with the target nuclei.

Since scattering in the absorbers and in the target is neglected, this means that the particles will retain their original directions of motion. The actual position of the various absorbers will therefore have no effect on the final output. The only parameter that remains is the total thickness of the absorbers expressed in g/cm². The geometric dimensions of the scatterers (if they are limited by constructional components) can be taken into account in the same way as the diaphragms. Figure 2 shows the diaphragm (detector) geometry. It will be shown later how the presence of a number of such diaphragms can be taken into account. For the moment, let us consider the simplest case: the telescope is characterized by a single recording area, and counts charged particles between R_1 and $R_2 = R_1 + \Delta R$, where ΔR is the range interval of the telescope, and R_1 represents the total amount of matter up to the detector.

Consider a volume element $dV = dx\,dy\,dz$ at an arbitrary point in the active volume of the target (with coordinates x, y, z). The number of nuclei in this volume element is $dn_n = (N_0 \rho / A) dV$, where N_0 is the Avogadro number, ρ is the density in g/cm³, and A is the atomic weight of the target material. We shall use the symbol $F(k, x, y, z)$ to denote the spatial and energy spectrum of the incident particles. The dimensions of this function are: particles/MeV·cm².

The number of secondary particles produced in the volume element dV is proportional to the number of incident particles per unit area, and to the total number of nuclei in dV, the proportionality coefficient being the cross section for the particular reaction. Therefore,

$$dN' = \sigma(k, \theta) F(k, x, y, z) dk \frac{N_0 \rho}{A} dV, \qquad (1.1)$$

where $F(k, x, y, z)dk$ is the number of primary particles with energies between k and k + dk which arrive on a unit area at the point x, y, z in the target. If the cross section is interpreted as the differential cross section, dN' is the number of secondary particles generated in the volume element dV and leaving it within a unit solid angle at an arbitrary angle θ to the beam direction.

Consider an area element $dS' = dy_i dz_i$ at the point $(L_i y_i z_i)$ in the plane of the detector. The number of particles reaching this area is proportional to the solid angle which it subtends at the point x, y, z. The angle θ is determined by the two straight lines $[(xyz), (L_i y_i z_i)]$ and the direction $\gamma\gamma$ of the beam. The solid-angle element is given to a good approximation by the formula

$$d\Omega = \frac{\cos\alpha\, dS}{l^2} = \frac{\cos\alpha\, dy_i\, dz_i}{l^2}, \qquad (1.2)$$

where α is the angle between the line $[(xyz), (L_i y_i z_i)]$ and the x axis, and l is the distance between the points (xyz) and $(L_i y_i z_i)$. It is clear that α and l are functions of both the coordinates of points in the target and of the coordinates of points in the plane of the detector. Therefore, the number of particles reaching an element of the detector is given by

$$dN = \sigma(k, \theta)\, F(k, x, y, z)\, dk\, \frac{N_0 \rho}{A}\, dV\, \frac{\cos\alpha}{l^2}\, dy_i\, dz_i. \qquad (1.3)$$

To obtain the total output, let us integrate this expression with respect to the incident-particle energy over the active volume of the target and over the detector area. The following condition must be then satisfied: we shall select only those cases where the recorded particles enter the energy interval of the telescope. This condition is reflected by the choice of the limits of integration.

Let us draw an arbitrary plane through the points (xyz) and $(L_i y_i z_i)$ which is parallel to the y axis, and consider an infinitely thin layer of material of thickness dz. We shall assume for simplicity that $z_i = z = a$ (Fig. 3). Here, $Q(x, y, z=a) = 0$ is the cross section of the target in the $z = a$ plane, and $q(x, y, z = a)$ is the beam cross section. The active area of this cross section is shown hatched.

For binary reactions and given incident-particle energy (and given masses), there is a kinematic relation between the energy and the angle of emission of the secondary particle:

$$E = f(k, \theta), \qquad (1.4)$$

where E is the energy of the secondary particle.

Using the relation

$$R(E) = \int_0^E \frac{dE}{-\frac{1}{\rho}\frac{dE}{dx}}, \qquad (1.5)$$

we can calculate the range in the target material [2, 3]. Let us draw from the point $(L_i, y_i, z_i = a)$ a bundle of rays covering the entire target. Each ray will be characterized by the angle θ_j with the direction of the incident particles, i.e., for each chosen direction (at the given energy k) we can calculate the range $R_j(k, \theta)$ in g/cm². If now from this range we subtract the lower limit of the range interval of the telescope referred to the target material, we obtain the maximum range in the target:

$$R_{mj} = R_j - R_1/\cos\alpha_j.$$

By laying off the quantities R_{mj}/ρ from the target outline in the inward direction, we obtain the upper boundary of the region from which the secondary particles can still reach the

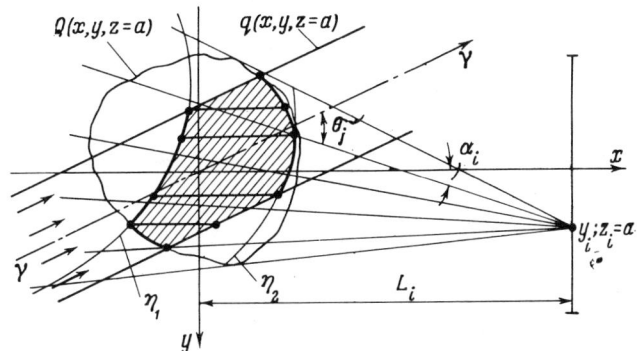

Fig. 3. Definition of the active volume of the target.

detector (the curve marked η_1). Similarly, the lower boundary (η_2) can be constructed from the upper boundary of the range interval of the telescope.

In the general case, the region of "active" target nuclei obtained in this way (shown hatched in Fig. 3) has a complicated contour which is determined by the target and beam geometry (active volume) and the bounding curves η_1 and η_2 which depend on the range interval of the telescope, the energy of the incident particles, the position of the area element dS, and the kinematics of the reaction under investigation. Having established the contour of the active region, we can find the limits of integration with respect to the variables x and y.

Thus, if we assume that the first integration is carried out with respect to x, it is clear from Fig. 3 that there is a number of regions (in the present case five) with different limits of integration: the first region extends from Q(x, y, z = a) to q(x, y, z = a), the second extends from η_1 to q(x, y, z = a), the third from η_1 to η_2, and so on. These limits are complicated functions of all the variables, i.e., y, z, y_i, z_i, k.

The limits of integration with respect to z are defined simply by the geometry of the active volume if the energy of the incident particles is independent of z. Unless this is so, they will be functions of k. The limits of integration with respect to y_i and z_i are determined by the detector contour, and the limits with respect to k can be found from the kinematic relation k = k(E, θ).

The reaction kinematics is usually such that k decreases with decreasing E and θ. Therefore, the lower limit of k can be determined from θ_{min} and E_{min} (and, correspondingly, R_1), and the upper limit can be determined from θ_{max} and $R_{max} = R_1 + \Delta R + R_m^{max}$, where R_m^{max} is the maximum range in the target. However, it is then necessary to take into account the specific geometry of the target and the possible dependence of R_m on θ.

Bearing in mind the above discussion on the limits and Eq. (1.3), we find that the output is given by the following expression:

$$N = \frac{N_0 \rho}{A} \int_{k_1}^{k_2} dk \int_{z_1(k)}^{z_2(k)} dz \int_{y_{1i}}^{y_{2i}} dy_i \int_{z_{1i}(y_i)}^{z_{2i}(y_i)} dz_i \int_{y_1(kzy_i z_i)}^{y_2(kzy_i z_i)} dy \sum_p \int_{x_{1p}(kyzy_i z_i)}^{x_{2p}(kyzy_i z_i)} \sigma(k,\theta) F(k, x, y, z) \frac{\cos \alpha}{l^2} dx, \quad (1.6)$$

where the subscript p represents summation over regions with different limits of integration.

If we take out the unknown cross section from under the integral sign, regarding it as a mean over the range of variation of the variables, we finally have

$$N = \frac{N_0 \rho}{A} \bar{\sigma} I \quad \text{or} \quad \bar{\sigma} = \frac{A}{N_0 \rho} \frac{N}{I}, \quad (1.7)$$

where I is the above multiple integral.

We have already noted that it is quite difficult to evaluate the integral I because this involves the determination of the limits of integration with respect to the spatial coordinates, which have a complicated energy dependence. In fact, when the functions η_1 and η_2 are calculated, and the points defining the various regions of integrations with respect to x are found, it is necessary to use a numerical method of calculation because the corresponding integral for the ranges cannot be evaluated in terms of quadratures. An alternative procedure is to use approximate range-energy relations for the various energy intervals. However, even in this case, the determination of the points of intersection of the curves η_1 and η_2 with the target contour involves the numerical solution of a set of nonlinear equations. This lengthens the time necessary for the calculations, complicates the programming logic, and hence increases the probability of errors in the computer realization of the given algorithm.

It is, however, possible to simplify the problem to some extent. Let us introduce an auxiliary function g_R into the integrand. By definition,

$$g_R = \begin{cases} 1, & \text{if } R_1 \leqslant R_x \leqslant R_1 + \Delta R, \\ 0, & \text{if } R_x < R_1 \text{ or } R_x > R_1 + \Delta R, \end{cases} \quad (1.8)$$

where R_x is the range of the particle after it leaves the target. The conditions given by Eq. (1.8) must be verified when the integration is carried out for all points of the target and detector.

Let us take an arbitrary point (xyz) in the active volume of the target, and an arbitrary point $(L_i y_i z_i)$ in the plane of the detector. The straight line $[(xyz), (L_i y_i z_i)]$ specifies a direction, i.e., we can define the angle θ to the beam direction and, for a given incident energy k, the energy and range of the emitted particle. If we now solve simultaneously the equations for the target surface $Q(x, y, z) = 0$ and the straight line $[(xyz), (L_i y_i z_i)]$ we obtain the point of intersection $(x_0 y_0 z_0)$ and hence the range of the particle in the target material. If $R(k, \theta)$ is the range of the particle created at the point (xyz), and $R_m(x, y, z, L_i, y_i, z_i)$ is its range in the target, then $R_x = R - R_m$ is its range after it leaves the target. The condition given by Eq. (1.8) must be verified for this quantity, thereby calculating the function g_R itself. The limits with respect to the space coordinates in the target then no longer depend on the energy, and are determined solely by the geometry of the active volume of the target, i.e., the problem is considerably simpler. However, it is then necessary to consider all the points in the active volume.

There are two opposing tendencies: on the one hand, we have reduced the time necessary for the calculations, since we no longer have to determine the exact boundaries of the active volume but, on the other, we have increased the time necessary to analyze (1.8) and consider the nonactive parts of the target. The gain in time as compared with the initial algorithm depends on each particular case. If the actual volume is not too different from the active volume there is a considerable gain in time. On the other hand, when the actual volume is much smaller than the active volume there may actually be a loss of time. Therefore, the optimal variant is to use both methods, i.e., at first, estimate the boundary of the actual volume by the first method, but only approximately, and then obtain the solution with the aid of the function g_R.

Let us return now to the diaphragms. The presence of one or more diaphragms means that for each point in the target there is a corresponding detector area. Conversely, for each point in the detector the actual volume of the target is additionally limited by diaphragm projections.

Here again, there are two methods: (a) we can try to describe the additional limitation of the volume by the combined geometry of the diaphragms, and hence improve the limits of integration with respect to z and y, and (b) we can introduce a further function g_s by analogy with g_R which would take the presence of diaphragms into account. The latter method is preferable.

By definition,

$$g_S = \Pi_i g_{S_i},$$

where

$$g_{S_i} = \begin{cases} 1, & \text{if the particle enters the contour, of the i-th diaphragm,} \\ 0, & \text{if the particle does not enter the contour of the i-th diaphragm.} \end{cases} \quad (1.9)$$

In this case, the integration is carried out over the area of the detector, and conditions (1.8) and (1.9) are simultaneously verified for each variable. This can be done quite simply because it is assumed that the functions $\varphi_i(y_i z_i) = 0$, which describe the diaphragm contours, are known.

We thus have

$$I = \int_{k_1}^{k_2} dk \int_V dV \int_{S_d} F(k, x\, y, z) \frac{\cos \alpha}{l^2} g_R g_S dS_d, \quad (1.10)$$

where V represents integration over the active volume of the target, and S_d represents integration over the detector area. The limits of integration are chosen on the basis of the geometry of the active volume and of the detector contour (subject to the condition that k is independent of the variables x, y, z, but if this is not so, then it must be allowed for in the determination of the limits with respect to these particular variables). The function g^R represents the energy discrimination of the telescope, and the function g_S the presence of the diaphragms.

§3. Multiple Scattering

The multiple scattering problem has been extensively investigated. Nevertheless, a complete and final theory is still not available. There are two main problems: (1) calculation of the differential scattering probability for an interaction between a charged particle and an atom in the target, and (2) calculation of the particle distribution function after a given layer of material.

If the nucleus can be regarded as a point charge, and the effect of atomic electrons can be ignored, the differential probability is given by the classical Rutherford formula

$$f(\theta) = \frac{z^2 Z^2 e^4}{4(c\beta p)^2 \sin^4 \theta/2}, \quad (1.11)$$

where β and p are the velocity and momentum of the scattered particle, ze is its charge, and θ is the scattering angle. In the small-angle approximation ($\sin \theta \approx \theta$) we have

$$f(\theta) = \frac{4 z^2 Z^2 e^4}{(c\beta p)^2} \frac{1}{\theta^4}. \quad (1.12)$$

The screening of the nucleus by the atomic electrons, and the effect of the finite size of the nucleus, can be determined as follows [4]. The scattering angle is $\theta \approx p'/p$, where p' is the transferred momentum at right angles to the direction of motion of the particle, and is inversely proportional to the impact parameter d. Therefore,

$$\theta \approx \frac{1}{dp} = \frac{\hbar}{d\hbar} \approx \frac{\lambdabar}{d},$$

where $\lambdabar$ is the de Broglie wavelength of the incident particle. If now instead of the impact parameter d we substitute the radius of the atom $a = a_0 Z^{-1/3}$, where a_0 is the first Bohr radius of hydrogen, and $b = 0.49 r_e A^{1/3}$ is the radius of the nucleus, then we obtain the upper and lower limits for the possible angles of scattering, i.e.,

$$\theta_{min} \approx \frac{\lambdabar}{a} \text{ and } \theta_{max} \approx \frac{\lambdabar}{b}. \tag{1.13}$$

Screening of the nucleus by the electrons is thus important for distances of the order of a, i.e., for scattering angles of about θ_{min}. This limits the range of application of Eq. (1.11). A similar analysis can be given for the upper limit of the scattering angles but, in this case, it is necessary to take into account the charge distribution in the nucleus. This estimate is, however, very approximate. Exact solution of the problem requires the use of the "true" atomic-field potential which would take into account both effects. Some workers [5-7] have used the Thomas–Fermi potential, but this does not yield an analytic expression for the differential scattering probability, and only numerical results can be obtained. Moreover, the Thomas–Fermi potential does not take into account the finite size of the nucleus, or the charge distribution, and describes satisfactorily only the screening in heavy atoms (high Z).

Other workers [8-11] have taken screening into effect by using a potential of the form $V(r) = \frac{zZe^2}{r} e^{-r/a}$. The differential scattering probability is then given by

$$f(\theta) = R [1 + (\lambdabar/2a \sin \theta/2)^2]^{-2}, \tag{1.14}$$

where R is given by the Rutherford formula (1.11).

The finite size of the nucleus was taken into account in [8]. The nucleus was regarded as a sphere of radius b in which the charge distribution was uniform. The final potential (including the screening effect) was

$$V(r) = \frac{zZe^2}{r} e^{-r/a} (1 - e^{-2r/b})$$

and the differential scattering probability turned out to be

$$f(\theta) = R \{[1 + (\lambdabar/2a \sin \theta/2)^2]^{-1} - [1 + (\lambdabar(a^{-1} + 2b^{-1})/2 \sin \theta/2)^2]^{-1}\}^2. \tag{1.15}$$

These expressions were obtained by solving the nonrelativistic Schroedinger equation using the Born approximation. A more rigorous solution based on the second Born approximation and allowing for relativity, the spin of the particles, and a screening potential of the form $V(r) = -\frac{zZe^2}{r} e^{-\lambda r}$ has been given by Dalitz [12]:

$$f(\theta) = \frac{4\alpha^2 \hbar^2 E^2}{c^2 [\hbar^2 \lambda^2 + 4p^2 \sin^2 \theta/2]^2} \{[1 - \beta^2 \sin^2 \theta/2] [1 - (\hbar^2 \lambda^2 + 4p^2 \sin^2 \theta/2) \times$$

$$\times \frac{\alpha}{\beta} \frac{1}{\pi^2} \operatorname{Re}(I+J) \Big] - (1-\beta^2)[\hbar^2\lambda^2 + 4p^2\sin^2\theta/2]\frac{\alpha}{\beta}\frac{1}{\pi^2}\operatorname{Re}(I-J) \Big\}, \tag{1.16}$$

where

$$\frac{1}{\pi^2}\operatorname{Re}(I \pm J) = -\frac{1}{\sin\theta/2\,\{\hbar^4\lambda^4 + 4p^2[\hbar^2\lambda^2 + p^2\sin^2\theta/2]\}^{1/2}} \times$$

$$\times \left[1 \pm \frac{\hbar^2\lambda^2 + 2p^2}{2p^2\cos^2\theta/2}\right]\tan^{-1}\left(\frac{\hbar\lambda p\sin\theta/2}{\{\hbar^4\lambda^4 + 4p^2[\hbar^2\lambda^2 + p^2\sin^2\theta/2]\}^{1/2}}\right) \pm$$

$$\pm \frac{1}{2p^2\cos^2\theta/2}\left[\tan^{-1}\left(\frac{2p}{\hbar\lambda}\right) - \frac{1}{\sin\theta/2}\tan^{-1}\left(\frac{p\sin\theta/2}{\hbar\lambda}\right)\right].$$

In these expressions p and E are the momentum and energy of the scattered particle (spin 1/2), λ is the screening parameter, $\alpha = zZe^2/\hbar c = zZ/137$ and $\beta = v/c$. The size of the nucleus was not taken into account in this formula, i.e., the nucleus was regarded as a point charge.

It was shown in [13] that the finite size of the nucleus can be obtained from the differential scattering cross section for a point charge by multiplying it by the nuclear formfactor which can be calculated from a model of the charge distribution. The model is usually chosen on the basis of electron scattering data. It is thus clear that there is a number of approximate solutions, and when attempts are made to obtain more accurate solutions, the corresponding expressions become exceedingly complicated. In some cases only numerical solutions are possible.

A similar situation occurs for the distribution function. In the simplest approach this problem reduces to the solution of the diffusion equation (the Fermi solution) which is given in [4, 14]:

$$P(x, y, \theta_y) = \frac{2\sqrt{3}}{\pi}\frac{1}{\theta_s^2 x}\exp\left[-\frac{4}{\theta_s^2}\left(\frac{\theta_y^2}{x} - \frac{3y\theta_y}{x^2} + \frac{3y^2}{x^3}\right)\right]. \tag{1.17}$$

In this expression x is the thickness of the target (the x axis lies along the initial direction of the particle), y is the lateral displacement at the exit from the scatterer in the xy plane, and θ_y is the projected angle of escape of the particle on the (x, y) plane; θ_s^2 is the mean square angle of scattering (it will be discussed in greater detail below).

The distribution function P is the probability density that the particle will undergo a lateral displacement y and scattering through an angle θ_y after passing through a thickness x of the material. By integrating the expression for P over all the possible values of y we obtain the angular distribution function

$$Q(x, \theta_y) = \int_{-\infty}^{+\infty} P(x, y, \theta_y)\,dy = \frac{1}{\sqrt{\pi}}\frac{1}{\theta_s x^{1/2}}\exp\left[-\frac{\theta_y^2}{\theta_s^2 x}\right], \tag{1.18}$$

i.e., the result is a Gaussian distribution. These are approximate results and are valid only for small-angle scattering. For larger angles the multiple scattering probability (1.18) becomes smaller than the probability given by the differential scattering cross section [see, for example, Eq. (1.11)].

This approximation arises in the derivation of the diffusion equation where it is assumed that the differential probability is nonzero in a very narrow range of angles, so that when the distribution function P is expanded into a series it is sufficient to retain only the first terms of the expansion. In the more general case, it is necessary to solve a complicated integro-differential equation. This type of analysis has been carried out in [5, 6, 9, 10] and elsewhere. In [9, 10] the final results were obtained only in numerical form. The final result, i.e., the distribution function, is given in [5, 7] in the form of a three-term analytic formula in which the first term gives the Gaussian distribution, the second gives the differential probability at large angles, and the third is a correction. Using the notation given in [7], this solution is of the form

$$f(\theta)\,\theta\,d\theta \approx v\,dv\,[f^{(0)}(v) + B^{-1}f^{(1)}(v) + B^{-2}f^{(2)}(v)], \tag{1.19}$$

where θ is the polar angle of deflection, $v = \theta/(\chi_c B^{1/2})$, and B is the solution of the equation $B - \log B = b$.

In these expressions

$$b \equiv \ln(\chi_c/\chi_a')^2 = \ln(\chi_c/\chi_a)^2 + 1 - 2C, \tag{1.20}$$

C is Euler's constant, and χ_a is the screening angle defined by

$$-\ln \chi_a = \lim_{k \to \infty} \left[\int_0^k q(\chi)\,d\chi/\chi - 1/2 - \ln k \right]. \tag{1.21}$$

Here, $q(\chi)$ is the ratio of the actual differential scattering cross section to the Rutherford cross section, and χ_c is a fictitious angle obtained from the Rutherford function on the assumption that the scattering probability through angles greater than or equal to χ_c throughout a foil of thickness t is equal to unity:

$$\chi_c^2 = 4\pi N t e^4 z^2 (Z+1)^2/(pv)^2. \tag{1.22}$$

The meaning of the other symbols is as follows: N is the number of nuclei per unit area, Z is the atomic number of the scattering material, and p, v, ze are the momentum, velocity, and charge of the scattered particle.

In his calculations, Moliere used the Thomas–Fermi potential to calculate the scattering cross section. Within the framework of this approximation the expression for b is

$$b = \ln \frac{6680 t}{\beta^2} \frac{(Z+1) Z^{1/3} z^2}{A(1+3.34\,\alpha^2)}, \tag{1.23}$$

where $\beta = v/c$, $\alpha = Zze^2/\hbar v$, and v is the velocity of the particle.

The expressions for the functions $f^{(i)}$ are as follows:

$$f^{(0)} = 2e^{-x}, \tag{1.24}$$

$$f^{(1)} = 2e^{-x}(x-1)[\bar{E}i(x) - \ln x] - 2(1 - 2e^{-x}), \tag{1.25}$$

$$1/4\, e^x f^{(2)} = [\psi^2(2) + \psi'(2)](x^2 - 4x + 2) + \int_0^1 t^{-3} dt\,[\ln t/(1-t) - \psi(2)] \times$$
$$\times [(1-t)^2 e^{xt} - 1 - (x-2)t - (1/2\,x^2 - 2x + 1)t^2], \tag{1.26}$$

where $x = v^2$, $\bar{E}i(x)$, $\psi(n)$, and $\psi'(n)$ are certain special functions tabulated in [15].

The integral in (1.26) can be evaluated with the aid of the power series given in [7]:

$$\int_0^1 = \sum_{n=0}^{\infty} \frac{1}{n+1} [\psi(n) + C - \psi(2)] \left[\frac{x^{n+3}}{(n+3)!} - \frac{2x^{n+2}}{(n+2)!} + \frac{x^{n+1}}{(n+1)!} \right] \quad (1.27)$$

and this can be used right up to $v = 10$ ($x = 100$).

Figure 4 shows plots of $f^{(0)}$ and $f^{(0)} + B^{-1} f^{(1)} + B^{-2} f^{(2)}$, calculated from the table given in [7]. In this figure $B = 10$, which corresponds to a target thickness of about 1 g/cm² of aluminum.

It is clear from Fig. 4 that an appreciable discrepancy occurs for $v \geq 2$, i.e., in the angular range for which the probability of deflection is about 0.01. Therefore, for all cases considered in practice, we can confine our attention to the first term of the expansion, i.e., to the Gaussian distribution of the form given by Eqs. (1.24) or (1.18).

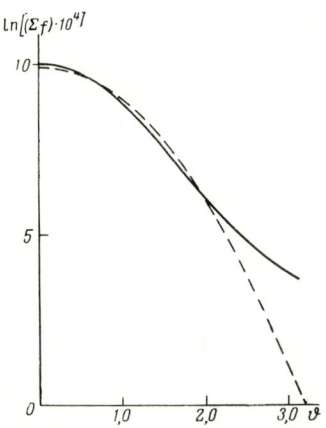

Fig. 4. Comparison of the Gaussian distribution function with the Moliere solution. Solid line: $f^{(0)} + B^{-1} f^{(1)} + B^{-2} f^{(2)}$; broken line: $f^{(0)}$.

As already noted, Eq. (1.18) was obtained by solving the diffusion equation

$$\frac{\partial P}{\partial t} = - \frac{\partial P}{\partial y} + \frac{1}{W^2} \frac{\partial^2 P}{\partial \theta_y^2}, \quad (1.28)$$

where

$$\frac{1}{W^2} = \frac{1}{2} \int \theta_y^2 d\theta_y \int f[(\theta_y^2 + \theta_z^2)^{1/2}] d\theta_z = \frac{\theta_s^2}{4} = \frac{1}{4} \int \theta^2 f(\theta) d\Omega, \quad (1.29)$$

$\theta^2 \approx \theta_y^2 + \theta_z^2$, $f(\theta)$ is the differential scattering probability, and θ_s^2 is the mean square of the scattering angle.

It was assumed in the derivation of (1.18) that W was independent of the energy of the incident particle, i.e., we neglected energy losses in the target.

The energy losses can be taken into account as shown in [16, 17] where, in the latter paper, it was noted that some of the final formulas in [16] were incorrect (probably due to typographic error). The distribution function given in [17] is

$$P(t, y, \theta_y) = (4\pi \sqrt{B})^{-1} \exp[-(\theta_y^2 A_2 - 2y\theta_y A_1 + y^2 A_0)/4B], \quad (1.30)$$

where

$$A_i = \int_0^t \frac{(t-\tau)^i}{W^2(\tau)} d\tau, \quad i = 0, 1, 2, \quad (1.31)$$

and $B = A_0 A_2 - A_1^2$.

The angular distribution function is

$$G_y(t, \theta_y) = \int_{-\infty}^{+\infty} P(t, y, \theta_y) dy = (2\sqrt{\pi A_0})^{-1} \exp[-\theta_y^2/4A_0]. \quad (1.32)$$

TABLE 1

Formula	Z		
	1	27	83
(1.37)/(1.36)	0.952	0.942	0.932
(1.38)/(1.36)	0.970	0.964	0.958

From Eqs. (1.29) and (1.31) it is clear that

$$A_0 = \int_0^t \frac{d\tau}{W(\tau)} = 1/4 \int_0^t \theta_s^2(\tau) d\tau = 1/4 \langle \theta^2 \rangle_{av}. \quad (1.33)$$

where $\langle \theta^2 \rangle_{av}$ is the mean square of the multiple-scattering angle calculated with allowance for energy losses since, by definition,

$$\theta_s^2 = \frac{d \langle \theta^2 \rangle_{av}}{dt} = \int \theta^2 f(\theta) d\Omega. \quad (1.34)$$

If we now substitute Eq. (1.33) in the expression for the angular distribution function given by Eq. (1.32), we obtain

$$G_y(t, \theta_y) = (\sqrt{\pi \langle \theta^2 \rangle_{av}})^{-1} \exp[-\theta_y^2 / \langle \theta^2 \rangle_{av}]. \quad (1.35)$$

To calculate θ_s^2 we must use the differential scattering probability. In view of the foregoing discussion there are several approximate expressions for θ_s^2:

1. The scattering probability is described by the Rutherford formula but only within the limits $\theta_1 = \theta_{min} \approx$ and $\theta_2 = \theta_{max}$, in which are estimated by Eq. (1.13):

$$\theta_s^2 = 4\pi N_0 \frac{Z(Z+1)}{A} r_e^2 \left(\frac{m_e c}{\beta p}\right)^2 \ln\left(\frac{\theta_2}{\theta_1}\right)^2. \quad (1.36)$$

2. The scattering probability is given by Eq. (1.14) in the angular range $0 \le \theta \le \theta_2$ and is zero for $\theta > \theta_2$:

$$\theta_s^2 = 4\pi N_0 \frac{Z(Z+1)}{A} r_e^2 \left(\frac{m_e c}{\beta p}\right)^2 \left\{\ln\left[\left(\frac{\theta_2}{\theta_1}\right)^2 + 1\right] - 1\right\}. \quad (1.37)$$

3. The scattering probability is given by Eq. (1.15):

$$\theta_s^2 = 4\pi N_0 \frac{Z(Z+1)}{A} r_e^2 \left(\frac{m_e c}{\beta p}\right)^2 \left\{\frac{4\theta_2^2 + \theta_1^2}{4\theta_2^2 - \theta_1^2} \ln \frac{4\theta_2^2}{\theta_1^2} - 2\right\}. \quad (1.38)$$

These calculations are carried out in the small-angle approximation, i.e., with $\sin \theta \approx \theta$.

Table 1 shows the ratio of Eqs. (1.37) and (1.38) to (1.36) for three values of Z.

It is clear from Table 1 that the difference between these solutions is quite small, and does not exceed 5% for moderate values of Z. In the ensuing analysis we shall use the expression given by Eq. (1.38). It can be simplified somewhat if we remember that

$$\left(\frac{\theta_2}{\theta_1}\right)^2 \approx \frac{280^2 \cdot 137^2}{(ZA)^{2/3}} \gg 1$$

even for very heavy elements. The factor in front of the logarithm is therefore approximately equal to unity, and we have the final expression

$$\theta_s^2 = 8\pi N_0 \frac{Z(Z+1)}{A} z^2 r_e^2 \left(\frac{m_e c}{\beta p}\right)^2 \left\{\ln \frac{7{,}66 \cdot 10^4}{(ZA)^{1/3}} - 1\right\}, \quad (1.39)$$

where N_0 is Avogadro's number.

The factor $Z+1$ (replacing Z) represents additional scattering by atomic electrons. The quantities z, β, and p are the charge, velocity, and momentum of the scattered particle, r_e is the classical radius of the electron, and m_e is its mass. Substituting numerical values for the constants, we obtain Eq. (1.39) in the form

$$\theta_s^2 = 0.315\, z^2 \frac{Z(Z+1)}{A} \frac{1}{(pv)^2} \left\{ \ln \frac{7.66 \cdot 10^4}{(ZA)^{1/3}} - 1 \right\}. \tag{1.40}$$

The quantity θ_s^2 in this formula is expressed in g/cm^2, and pv is measured in MeV.

We thus arrive at the following conclusions:

1. In practical calculations of the output with allowance for multiple scattering in the target and the telescope absorbers we can use the Gaussian expression for the distribution function.

2. The geometric thickness of the absorbers is usually much smaller than the distance between them, and therefore we can ignore the lateral displacement at the exit from the absorber, and consider only the angular divergence of the scattered particles.

3. The different degrees of approximation used for the differential scattering probability have little effect on the mean square angle of scattering, which governs the variance of the distribution function.

4. When the energy losses in an absorber are considerable (and an absorber of this kind can always be found in our detection system), the mean square multiple-scattering angle must be calculated with allowance for these losses.

Expression for the Output with Allowance for Multiple Scattering. While in the derivation of the equation for the output without allowance for multiple scattering we could confine our attention to a relatively narrow angular range determined by the geometry of the target and telescope, in the case where multiple scattering is included the direction is still conserved but only in the probabilistic sense, namely, the probability maximum still lies in the original direction of the particle, all other directions become possible, at least in principle. Therefore, if we take scattering in the target into account we must consider all the angles of escape from the target. The initial angular parameters describing the motion of the particle inside the target are restricted only by the kinematics of the process.

We shall now write down the angular distribution function for the polar angle of deflection. The expression given by Eq. (1.35) was derived for the projected angle in the (xy) plane (it was assumed that the initial direction of the particle lay along the x axis). The probability that the particles will leave in the angular range θ_y, $\theta_y + d\theta_y$ after traversing a layer of thickness R is

$$dP_y = G_y(R, \theta_y)\, d\theta_y = \frac{1}{\sqrt{\pi \langle \theta^2 \rangle_{av}}} \exp\left[-\frac{\theta_y^2}{\langle \theta^2 \rangle_{av}}\right] d\theta_y.$$

In view of the symmetry of the situation we can write down the corresponding expression for the projected angle in the (xz) plane. It is clear that the probability of simultaneous realization of θ_y and θ_z is equal to the product of these probabilities. Therefore,

$$dP = \frac{1}{\pi \langle \theta^2 \rangle_{av}} \exp\left[-\frac{\theta^2}{\langle \theta^2 \rangle_{av}}\right] d\Omega = G(R, \theta)\, d\Omega, \tag{1.41}$$

where θ is the polar angle of deflection, and in the small-angle approximation $\theta^2 \simeq \theta_y^2 + \theta_z^2$, $d\Omega$ is the solid-angle element, and

$$\langle \theta^2 \rangle_{av} = \int_0^R \theta_s^2 \, dR, \qquad (1.41a)$$

where θ_s^2 is given by Eq. (1.40).

Figure 5 shows the geometry of target and telescope. For the sake of simplicity we show only the projection onto the x, y plane. The notation is the same as in Fig. 3.

This diagram shows only a number of diaphragms (D_i), scatters (R_i), and a detector DT. As before, we are assuming that the functions

$$\psi_{D_i}(y_i^*, Z_i^*) \text{ and } \psi_{R_i}(y_i, Z_i),$$

which describe the contours of the corresponding diaphragms and scatterers, and the contour of the detector are known.

Let us take an arbitrary point (xyz) in the active volume of the target, and consider a volume element dV = dxdydz around it. Moreover, in the plane of, for example, the first scatterer let us take an arbitrary element of area $dS_1' = dy_1' dz_1'$ whose center has the coordinates (L_1, y_1', z_1'). This defines the direction of the ray $[(xyz), (L_1 y_1' z_1')]$, and the angle θ with the direction $\gamma\gamma$ of the beam. Using Eq. (1.3), we can then write

$$dN_1 = \sigma(k, \theta) F(k, x, y, z) dk \frac{N_0 \rho}{A} dV \frac{\cos \alpha_1'}{l_1'^2} dy_1' dz_1', \qquad (1.42)$$

where l_1' is the length of the ray $[(x, y, z), (L, y_1', z_1')]$, and α_1' is the angle between this ray and the x axis. The quantity dN_1 is then the number of particles generated in the volume element dV as a result of interaction with the primary particles Fdk which have energies between k and k+dk and escape in the direction of the element dS_1' within a solid angle $dS_1' \cos \alpha_1'/l_1'^2$. As before, let us determine the coordinates $(x_0 y_0 z_0)$ of the point of intersection of the ray l_1' with the surface of the target, and calculate the possible range $R(k, \theta)$ of the secondary particle and its range R_0 in the target.

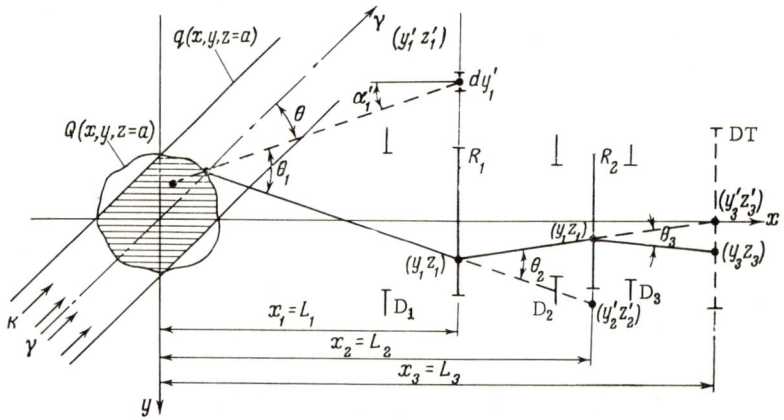

Fig. 5. Scheme for the telescope detection of a charged particle with allowance for scattering in the target and in absorbers.

As a result of multiple scattering over the path length R_0, the dN_1 particles leaving the target will have an angular distribution of the form given by (1.41). Let us take an arbitrary area element $dS_1 = dy_1 dz_1$ centered on the point (L, y_1, z_1) within the contour of the first scatterer, and thus specify the polar angle of deflection θ_1 from the initial direction. The number of particles entering the area element dS_1 is given by

$$dN_2 = dN_1 G_1(R_0 \theta_1) d\Omega_1 = dN_1 G_1(R_0 \theta_1) \frac{\cos \alpha_1}{l_1^2} dy_1 dz_1, \qquad (1.43)$$

where α_1 is the angle between the straight line $[(x_0 y_0 z_0), (L_1 y_1 z_1)]$ and the x axis, and l_1 is the length of the segment $[(x_0 y_0 z_0), L_1 y_1 z_1)]$. These particles will pass through the first scatterer (range $R_1/\cos \alpha_1$) and, as a result of multiple scattering, will again form a certain angular distribution at the exit.

Let us now take an arbitrary area element dS_2 in the plane of the second scatterer, and calculate the angle of deflection θ_2 and the probability of entering dS_2, and so on until we reach the detector.

In this way the first differential of the output can be shown to be

$$dN = \sigma(k, \theta) F(x, y, z, k) dk \frac{N_0 \rho}{A} dx\, dy\, dz \frac{\cos \alpha_1'}{l_1'^2} dy_1' dz_1' \prod_i G_i(R_{i-1}\theta_i) \frac{\cos \alpha_i}{l_i^2} dy_i dz_i. \qquad (1.44)$$

To obtain the total output we must integrate this equation. As before (when scattering was not taken into account), let us introduce the functions g_R and g_S, which represent the energy band of the telescope and the geometry of the diaphragms. Here we must note that the conditions for the function g_R will be somewhat different, namely,

$$g_R = \begin{cases} 1, & \text{if} \quad 0 \leqslant R_x \leqslant \Delta R, \text{ where } R_x = R - R_0 - \sum_i R_i/\cos\alpha. \\ 0, & \text{if} \quad R_x < 0 \text{ or } R_x > \Delta R. \end{cases}$$

The function g_S must be calculated in steps, i.e., transition to each new scatterer requires the verification of the condition that the corresponding diaphragm has been traversed.

Consequently, we have

$$N = \bar{\sigma} \frac{N_0 \rho}{A} \int_{k_1}^{k_2} dk \int_V F\, dV \int_{-\infty}^{+\infty} \frac{\cos \alpha_1'}{l_1'^2} dS_1' \int_{S_1} G_1 g_{S_1} dS_1 \ldots \int_{S_n = S_{dt}} G_n g_{S_n} g_R dS_n. \qquad (1.45)$$

The limits of integration within the volume of the target remain as before, i.e., they are determined by the geometry of the active volume. The limits of the integral with respect to the energy k are somewhat different. The lower limit remains the same as before, i.e., as in the case without scattering, since it is determined by the condition for the minimum angle and minimun range. The upper limit is higher, since now the maximum angle of escape is not restricted, and scattering occurs throughout the target. The upper limit may be restricted by kinematic conditions, or the maximum energy in the spectrum of the incident particles.

The limits in the integrals with respect to S_1', which define the initial direction of the particle, if there are no kinematic restrictions, can, in principle, be infinite. However, in practice, the distribution function falls relatively rapidly with increasing angle, and the limits can in fact always be taken to be finite. The limits of the other integrals are determined by the geometry of the scatterers and the detector.

We note that the inclusion of a large number of scatterers makes the calculations extremely laborious, since each scatterer adds two further integrals to the existing six integrals in the problem without scattering.

CHAPTER II

Calculation of the Output in the Case of the Compton Effect and the Photoproduction of Neutral Mesons on Hydrogen

§1. Definition and Addtional Restrictions

We shall now continue our analysis, applying it to specific experimental conditions.

Flux and Spectrum of Bremsstrahlung. The bremsstrahlung spectrum was calculated as in [18] with allowance for the energy distribution of primary electrons in the accelerator which is due to the "stretching" of the radiation intensity in time. According to [18], the probability that an electron with initial energy E_0 will emit a photon with energy between k and k+dk as a result of an interaction with a target nucleus in the synchrotron is given by

$$\sigma(E_0, k; \theta) \, dk \, d\theta = \frac{4Z^2}{137} \left(\frac{e^2}{mc^2}\right)^2 \frac{dk}{k} \theta \, d\theta \left\{ \frac{16\theta^2 E}{(\theta^2+1)^4 E_0} - \frac{(E_0+E)^2}{(\theta^2+1)^2 E_0^2} + \left[\frac{E_0^2+E^2}{(\theta^2+1)^2 E_0^2} - \frac{4\theta^2 E}{(\theta^2+1)^4 E_0}\right] \ln M(x) \right\}, \tag{2.1}$$

where θ is the angle of emission, and

$$\frac{1}{M(x)} = \left(\frac{\mu k}{2EE_0}\right)^2 + \left(\frac{Z^{1/3}}{C(\theta^2+1)}\right)^2. \tag{2.2}$$

In these expressions $E = E_0 - k$, $\mu = mc^2$, m is the mass of the electron, e is its charge, C is a dimensionless constant (C=111), and Z is the atomic number of the target.

The above expression is valid for very thin targets. However, if the experimental setup involves an angular divergence of the bremsstrahlung radiation which is of the order of mc^2/E_0 then, with good accuracy, the photon energy distribution is described by (2.1) integrated with respect to all angles even when the target is quite thick, i.e., when multiple scattering of the electrons takes place [19]:

$$\sigma(k, E_0) \, dk = \frac{2Z^2}{137} \left(\frac{e^2}{mc^2}\right)^2 \frac{dk}{k} \left\{ \left(\frac{E_0^2+E^2}{E_0^2} - \frac{2}{3}\frac{E}{E_0}\right) \times \right.$$

$$\left. \times \left(\ln M(0) + 1 - \frac{2}{b}\tan^{-1} b\right) + \frac{E}{E_0}\left[\frac{2}{b^2}\ln(1+b^2) + \frac{4(2-b^2)}{3b^3}\tan^{-1} b - \frac{8}{3b^2} + \frac{2}{9}\right] \right\}, \tag{2.3}$$

where $b = 2E_0 E Z^{1/3}/C\mu k$ and M(0) is determined from (2.2) with $\theta = 0$.

Substituting $E = E_0 - k$, and replacing E_0 with E, we obtain

$$\sigma(k, E) \, dk = B \frac{dk}{k} \left\{ \left[\frac{4}{3}\left(1 - \frac{k}{E}\right) + \left(\frac{k}{E}\right)^2\right]\left(\ln M(0) + 1 - \frac{2}{b}\tan^{-1} b\right) + \right.$$

$$\left. + \left(1 - \frac{k}{E}\right)\left[\frac{2}{b^2}\ln(1+b^2) + \frac{4(2-b^2)}{3b^3}\tan^{-1} b - \frac{8}{3b^2} + \frac{2}{9}\right] \right\}. \tag{2.4}$$

In these expressions B is a constant ($B = 2Z^2 r_e^2/137$). If the primary electrons have an energy distribution F(E) in the interval between E_1 and E_2, the number of photons with energy k, which are due to electrons with energies between E and E + dE, is given by

$$f'(k, E)dkdE = \sigma(k, E)dk F(E)dE, \qquad (2.5)$$

and the total number of photons with energies k due to the entire electron spectrum is obtained by integration with respect to E:

$$f(k, E_1, E_2)dk = dk \int_{E_1}^{E_2} \sigma(k, E) F(E) dE. \qquad (2.6)$$

When the bremsstrahlung flux is measured in practice, a determination is usually made of the energy flux, so that it is more convenient to use the normalized bremsstrahlung spectrum, i.e., the spectrum referred to the amount of energy carried by the photons in the entire spectrum. In our case, when the electrons in the accelerator have the distribution F(E) this normalization factor is defined as follows:

$$\Pi_0 = \int_0^{k_{max}} k \, dk \int_{E_1}^{E_2} \sigma(k, E) F(E) dE, \qquad (2.7)$$

where $k_{max} = E_2 - mc^2$ is the maximum photon energy.

For the normalized bremsstrahlung spectrum we then have the following expression:

$$f_n(k, E_1, E_2) = \frac{1}{\Pi_0} \int_{E_1}^{E_2} \sigma(k, E) F(E) dE = \frac{1}{k} \frac{\int_{E_1}^{E_2} \Phi(k, E) F(E) dE}{\int_0^{k_{max}} dk \int_{E_1}^{E_2} \Phi(k, E) F(E) dE}, \qquad (2.8)$$

where the function $\Phi(k, E)$ is the part of the bremsstrahlung cross section which is equal to the quantity in the curly brackets in Eq. (2.4). The dimensions of f_n are quantum/MeV^2.

If the flux measured during the time of the experiments is Π (MeV), the total number of photons with energies between k and k + dk which have passed through the apparatus (in the entire beam) is $\Pi f_n dk$.

Spatial Distribution of the Bremsstrahlung Flux. The bremsstrahlung flux density is not constant. It has an axial symmetry and falls quite rapidly with distance from the center. Figure 6 shows an example of the radial dependence of the flux density in the plane perpendicular to the beam axis. This distribution can be represented quite well by the exponential formula

$$\frac{d\Pi}{dS} \approx A e^{-r/r_0}. \qquad (2.9)$$

Only a part of the bremsstrahlung beam is usually separated out in the experiment (shown shaded). Collimators are set up along the beam axis, and the energy flux is measured after the collimator. If the flux measured in the time of the experiment is Π_1, the flux density is given by

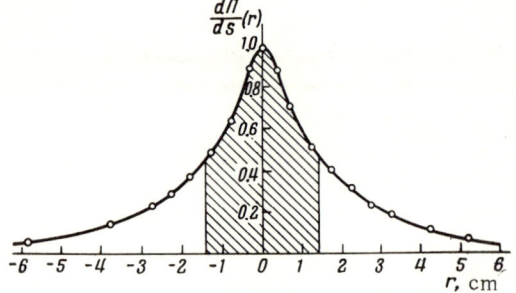

Fig. 6. Bremsstrahlung flux-density distribution. Open circles: experimental [20]; solid lines: e^{-r/r_0}, $r_0 = 1.9$ cm, $L = 184$ cm.

$$\frac{d\Pi}{dS} = \frac{\Pi_1 e^{-r/r_1}}{2\Pi r_0^2 [1 - e^{-r_1/r_0}(1 + r_1/r_0)]}. \qquad (2.10)$$

In the above expression r_1 is the beam radius, and r_0 is a parameter which can be deduced from the measured distribution density [20].

The slight divergence of the beam (about 1°) can be neglected, and it is therefore considered that the beam is defined by a cylindrical surface of radius r_1 in the region of the target. The intensity distribution is given by Eq. (2.10) and the proton spectrum by Eq. (2.8).

Target. The target is a thin, cylindrical container or radius R, located at right-angles to the beam axis, and filled with liquid hydrogen. The equation of the surface of the target in our coordinate system is

$$x^2 + y^2 = R^2. \qquad (2.11)$$

TABLE 2

Medium	a	b
Hydrogen	$7.636 \cdot 10^{-4}$	1.8366
Aluminum	$2.852 \cdot 10^{-3}$	1.7728

Telescope. The computational scheme for the telescope has already been considered. The number of absorbers (scatterers) is not restricted. It is assumed that not more than one diaphragm is placed in front of each absorber. The contours of the absorbers and diaphragms are coaxial and symmetric. Two types of contour are considered: a) rectangular and b) circular. The corresponding equations are

$$\text{a)} \quad z_i = \pm \rho_{z_i}, \qquad \text{b)} \quad y_i^2 + z_i^2 = \rho_i^2,$$
$$x_i = L_i, \qquad \qquad x_i = L_i.$$
$$y_i = \pm \rho_{y_i};$$

The Range — Energy Relation. In the reactions which we shall consider, the secondary particles are protons with energies of about 10-100 MeV and, therefore, the range energy relation in this range interval can be approximated to by the formula

$$R = aE^b \qquad (2.12)$$

which will be valid for the two media, namely, hydrogen (target) and aluminum (filters). The constants were determined by the method of least squares, using the points given in [2, 3]. The results are given in Table 2.

The deviation of ranges calculated in this way from standard points did not exceed 1%.

Mean Square Multiple-Scattering Angle. When the expression for θ_s^2 given by Eq. (1.40) is integrated, it is important to take into account the energy dependence of the form $A/(pv)^2$, where A is a constant which depends only on the material of the scatterer. Since, in our energy range, the protons are still nonrelativistic, the expression $1/(pv)^2$ can be expanded into a series in powers of E/m, where E is the kinetic energy and m is the proton mass, and retain only the linear terms of the expansion.

This gives the following expression

$$\frac{1}{(pv)^2} \approx \frac{1}{4E^2}\left(1 + \frac{E}{m}\right). \qquad (2.13)$$

Substituting this into Eq. (1.41a), and using Eq. (2.12), we obtain the expression for the mean square multiple-scattering angle:

$$\langle \theta^2 \rangle = \frac{0.315\, Z(Z+1)}{A}\left[\ln\frac{7.66\cdot 10^4}{(ZA)^{1/3}} - 1\right]\frac{b}{4}\left\{\frac{a^{2/b}}{b-2}\left(R_1^{\frac{b-2}{b}} - R_2^{\frac{b-2}{b}}\right) + \frac{a^{1/b}}{m(b-1)}\left(R_1^{\frac{b-1}{b}} - R_2^{\frac{b-1}{b}}\right)\right\}, \qquad (2.14)$$

where R_1 and R_2 are the range of the proton before and after scatterer, respectively, m is the proton mass, and a and b are the parameters in the approximation defined by Eq. (2.12).

§2. Evaluation of the Multiple Integral

It was shown at the end of Chapter I [see Eq. (1.45)] that the solution of our problem involves the evaluation of a multiple integral which even in the simplest case, for example, when only scattering in the target and one absorber is taken into account, is, in fact, an integral of multiplicity equal to 10. Such integrals cannot be evaluated by using uniform grids because the number of evaluations of the integrand is then N^n, where N is the number of steps for one of the variables (assumed the constant), and n is the mulitiplicity of the integral.

In the present case it is convenient to use the Monte-Carlo method [21-23].

Let us consider this problem in terms of the following example.

It is required to evaluate the definite integral

$$I = \int_a^b f(x)\, dx. \qquad (2.15)$$

Let us suppose that x is a random quantity which is distributed with a probability density p(x) in the interval (a-b). Moreover,

$$\int_a^b p(x)\, dx = 1.$$

Mathematical statistics then shows that the mathematical expectation of $f(x)$ is

$$Mf(x) = \int_a^b p(x) f(x)\, dx. \qquad (2.16)$$

If now the probability density p(x) is given a particular form, namely, the uniform distribution

$$p(x) = \begin{cases} \dfrac{1}{b-a} & \text{for} \quad a \leqslant x \leqslant b, \\ 0 & \text{for} \quad x < a \text{ or } x > b, \end{cases}$$

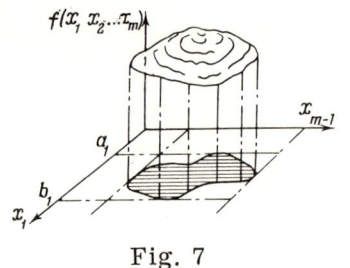

Fig. 7

we obtain

$$Mf(x) = \frac{1}{b-a} \int_a^b f(x)\,dx = \frac{1}{b-a} I \qquad (2.17)$$

or

$$I = (b-a)\,Mf(x).$$

Therefore, to evaluate the integral we must find an estimate of the mathematical expectation of the random quantity $y_i = f(x_i)$. It can be shown that a good estimate of the mathematical expectation is the arithmetic mean, i.e.,

$$Mf(x) \approx \frac{1}{N} \sum_{i=1}^{N} y_i = Y. \qquad (2.18)$$

To determine the error in the integral evaluated in this way we must estimate the variance of Y. Suppose that $D_{y_i} = D_y$ [21]. We then have

$$DY = \frac{1}{N^2} \sum_i Dy_i = \frac{Dy}{N} = \frac{Df(x)}{N}. \qquad (2.19)$$

The central limit theorem of the theory of probability then shows that, for a sufficiently high number N of tests, the random quantity Y is distributed approximately normally, and the error $\sigma_y = \sqrt{DY}$ corresponds approximately to the same level of expectation with which other physical quantities, for example, the output, are determined.

To evaluate the variance DY we must know the variance of the integrand $f(x)$, i.e.,

$$Df(x) = \frac{1}{b-a} \int_b^a f^2(x)\,dx - \frac{1}{(b-a)^2} I^2. \qquad (2.20)$$

This requires not only a knowledge of I, which we wish to evaluate, but also of the integral of the square of the integrand. Therefore, to estimate $Df(x)$ we shall use the approximate expression

$$Df(x) \approx \frac{1}{N} \sum_i y_i^2 - \left(\frac{1}{N} \sum_i y_i\right)^2, \qquad (2.21)$$

which ensures sufficient accuracy for large N.

In practice it is usual to employ a single standard set of random quantities which are distributed uniformly in the interval 0-1. The range of integration is therefore reduced to this interval by changing the variable. Thus, in our example, if we substitute $\zeta = (x-a)/(b-a)$ we obtain

$$I = (b-a) \int_0^1 f[a + (b-a)\,\xi]\,d\xi,$$

where ξ is a random quantity which is distributed uniformly in the interval 0-1.

We can now use standard methods to generate such random numbers. In the case of multiple integrals the problem is complicated by the fact that the limits of integration are, in general, related and form a multidimensional range of integration (Fig. 7). Let

$$I = \iint_s \cdots \int f(x_1, x_2, \ldots, x_m)\, dx_1 \ldots dx_m.$$

We shall now determine the maximum possible limits for each variable, i.e., $a_i \leq x_i \leq b_i$, $i = 1, 2, \ldots, m$, and introduce the new variables

$$x_i = a_i + (b_i - a_i)\, \xi_i. \tag{2.22a}$$

Consider the integral

$$I = \iint_\sigma \cdots \int F(\xi_1, \xi_2, \ldots, \xi_m)\, d\xi_1\, d\xi_2 \ldots d\xi_m, \tag{2.23}$$

where

$$F(\xi_1, \ldots, \xi_m) = \Pi_i (b_i - a_i)\, f\{[a_1 + (b_1 - a_1)\xi_1],\ [a_2 + (b_2 - a_2)\xi_2],\ \ldots,\ [a_m + (b_m - a_m)\xi_m]\}. \tag{2.24}$$

We have thus "inscribed" the range of integration into an m-dimensional unit cube.

Each statistical test now involves the selection of m independent and uniformly distributed numbers in the interval 0–1, i.e., ξ_{1l} ξ_{2l} …, ξ_{ml}, which specify a certain random point within the m-dimensional cube (the subscript l represents the number of the test). Two cases are possible, namely, (1) the random point M_l ($\xi_{1l}, \xi_{2l}, \ldots, \xi_{ml}$) lies in the region of integration i.e., $M_l \in \sigma$ and this is called a successful test, and (2) conversely, M_l does not lie in σ, i.e., $M_l \overline{\in} \sigma$ and this is defined as an unsuccessful test.

We shall assume that the total number of tests is N. Of these, n are successful. In that case we have

$$I \approx \sigma \frac{1}{n} \sum_l F_l(M_l). \tag{2.25}$$

If the size σ of the region is unknown, we can take approximately

$$\sigma \approx \frac{n}{N}. \tag{2.26}$$

In that case,

$$I \approx \left(\frac{1}{n} \sum_l F_l\right) \frac{n}{N} = \frac{1}{N} \sum_l F_l. \tag{2.27}$$

This approach is analogous to the extension of the range of integration to a broader interval but subject to the condition

$$F_l(M_l) = \begin{cases} F_l(M_l), & \text{if } M_l \in \sigma, \\ 0 & \text{if } M_l \overline{\in} \sigma, \end{cases} \tag{2.28}$$

which in turn is analogous to the introduction of an additional function g, such that

$$g = \begin{cases} 1, & \text{if } M_l \in \sigma, \\ 0 & \text{if } M_l \overline{\in} \sigma, \end{cases} \qquad (2.29)$$

and integration within the constant limits 0-1:

$$I = \int_0^1 \int_0^1 \ldots \int_0^1 F(\xi_1, \xi_2, \ldots, \xi_m) g \, d\xi_1 \, d\xi_2 \ldots d\xi_m. \qquad (2.30)$$

Estimated Accuracy of the Integral. It is clear from Eq. (2.27) that I is obtained in the form of a product of two random quantities, namely, $I = XY$, where $X = \frac{1}{n}\sum_l F_l$ is the estimated mathematical expectation of the integrand (distributed approximately uniformly), and $Y = n/N$ is the estimated range of integration which is a part of the m-dimensional unit cube and and is distributed in accordance with the binomial law.

In practical calculations we are mainly interested in the relative error in the integral. The square of the relative error (we shall call this the relative variance) is given by

$$\delta^2 I \equiv \frac{DI}{(MI)^2} = \frac{DX}{(MX)^2} + \frac{DY}{(MY)^2}, \qquad (2.31)$$

where DI, DX, and DY are the variances of the corresponding quantities, and MI, MX, and MY are their mathematical expectations. The quantities X and Y are independent. Let $MY = p$. In that case $DY = pq/N$, where $q = (1-p)$. We can use (2.19) and replace n by its mathematical expectation $Mn = Np$, so that

$$\delta^2 I = \frac{1}{Np}\left[\frac{DF}{(MF)^2} + (1-p)\right]. \qquad (2.32)$$

The first term in the square brackets represents the relative variance of the integrand. The second term is a measure of how well the region of integration is inscribed into the unit cube. For example, if the region is not adequately inscribed, i.e., p is very small ($p \to 0$), then it is practically impossible to obtain good relative accuracy. Special measures must then be taken, for example, the region of integration must be divided into individual regions, and the integral evaluated part by part.

Moreover, we must, of course, satisfy the main condition, namely, that the relative variance of the integrand must be finite and as small as possible. This condition is satisfied if the mathematical expectation of the integrand does not tend to zero, but if the integrand has singularities then it must be both integrable and square integrable. Let us consider the output integrand from this point of view.

§3. Analysis of the Output Integrand

Bearing in mind the remarks introduced in §1, we can rewrite the expression for the output in the form

$$B = \bar{\sigma}\frac{N_0\rho}{A}\int_{k_1}^{k_2}\int_V\int_{-\infty}^{\infty}\int_{S_1}\ldots\int_{S_n} f_n \frac{d\Pi}{ds}\frac{\cos\alpha_1'}{l_1'^2}\Pi_i\left(G_i(R_i\theta_i)\frac{\cos\alpha_i}{l_i^2}g_{s_i}\right)g_R \, dk \, dV \, dS_1' \, dS_1 \ldots dS_n. \qquad (2.33)$$

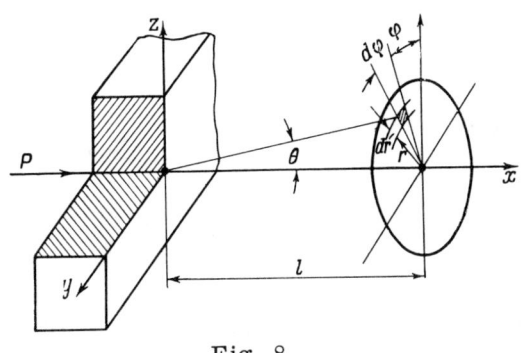

Fig. 8

Neither f_n nor $\cos \alpha_i / l_i^2$ have any singularities for $k \neq 0$.

Consider the function

$$G_i(R_i \theta_i) = \frac{1}{\pi \langle \theta^2 \rangle_i} \exp[-\theta_i^2 / \langle \theta^2 \rangle_i], \quad (2.34)$$

$$\langle \theta^2 \rangle_i = \int_0^{R_i} \theta_s^2 \, dR.$$

Since real scatterers have finite thickness, the functions G_i are finite throughout the range of the variables. However, if in addition to scatterers we include scattering in the target, then, in general (for example, when the entire volume of the target lies in the beam), it is necessary to consider even those points of the target which lie on its surface, i.e., where the thickness of the scatterer is zero. In this case, $\langle \theta^2 \rangle_i$ is zero, and for $\theta_i \to 0$ we have a divergence, i.e., $G_i(R_i \theta_i) \to \infty$ as $R_i \to 0$ and $\theta_i \to 0$.

To estimate the effect of this divergence on the variance of the integrand, let us consider the simple example shown in Fig. 8 where the particles move along the axis of the telescope, and after leaving the scatterer are recorded by a flat circular counter of radius R. In this case, it is simpler to transform to the cylindrical set of coordinates (r, φ) in the plane of the counter.

In the small-angle approximation ($\sin \theta \approx \tan \theta \approx \theta$) the probability that the particle will reach the area element $dS = r\, dr\, d\varphi$ is given by

$$dP = \frac{1}{\pi \langle \theta^2 \rangle} \exp\left[-\frac{r^2}{l^2 \langle \theta^2 \rangle}\right] \frac{r\, dr\, d\varphi}{l^2}, \quad (2.35)$$

where $\langle \theta^2 \rangle$ depends on the energy of the incident proton and the thickness of the filter. Let us introduce the new parameter $r_0 = \sqrt{l^2 \langle \theta^2 \rangle}$ and integrate (2.35) with respect to φ:

$$dP = \frac{2r}{r_0^2} \exp[-r^2 / r_0^2]\, dr. \quad (2.36)$$

To obtain the probability that the particle will enter the counter we must integrate Eq. (2.36) with respect to r:

$$P = \int_0^R f(r)\, dr = \int_0^R \frac{2r}{r_0^2} e^{-r^2/r_0^2}\, dr. \quad (2.37)$$

Let us now calculate the mathematical expectation and variance of the integrand on the assumption that the random quantity r is uniformly distributed in the interval $0 - R$:

$$Mf = \frac{1}{R} \int_0^R f(r)\, dr = \frac{1}{R}(1 - e^{-R^2/r_0^2}), \quad (2.38)$$

$$Mf^2 = \frac{1}{R} \int f^2(r)\, dr = \frac{\sqrt{2\pi}}{4Rr_0} \Phi\left(\frac{2R}{r_0}\right) - \frac{1}{r_0^2} e^{-\frac{2R^2}{r_0^2}}, \quad (2.39)$$

where Φ is the error integral.

The relative variance of the integrand is then given by

$$\delta^2 f = \frac{\frac{\sqrt{2\pi}}{4Rr_0}\Phi\left(\frac{2R}{r_0}\right) - \frac{1}{r_0^2}e^{-\frac{2R^2}{r_0^2}}}{\frac{1}{R^2}(1 - e^{-R^2/r_0^2})^2} - 1. \tag{2.40}$$

Let us now consider two limiting cases: $r_0 \to 0$ and $R \to \infty$.

The first case occurs in the analysis of target regions near the surface. It is clear from (2.40) that when $r_0 \to 0$ we have

$$(Mf)^2 \to 1/R^2, \ Df \to \infty, \text{ and } \delta^2 f \to \infty.$$

This conclusion remains if we consider not only the central direction of the incident particle but all other directions as well. Moreover, in the general expression for the output given by Eq. (2.33) the integrand is the product of separate functions and, therefore, in the first approximation, the relative variance of the integrand will consist of the sum of the relative variances of these functions. Divergent behavior of one of the terms will then ensure that the relative variance of the entire integrand will tend to infinity.

The second case is never realized in pure form because we must always integrate over a finite area of the scatterer, but it is of interest when $R_2 \gg r_0^2$ which may frequently occur in practice (for example, in the case of thin scatterers or relatively compact geometry).

It is clear from Eq. (2.40) that $(Mf)^2 \to 0$ (as $1/R^2$), $Df \to 0$ (as $1/R$, and $\delta^2 f \to \infty$ (approximately as R), when $R \to \infty$, i.e., although the variance decreases with increasing R/r_0 the relative accuracy will deteriorate. Consequently, the problem involving corrections for multiple scattering in the target cannot be solved on this scheme.

To avoid this difficulty let us restrict the range of existence of the differential scattering probability. Since this probability is an exponential function of r^2, we may assume that the main "mass" of the probability is concentrated within a region of radius $r' = nr_0$, where n is a positive number. Substituting $\xi = r/nr_0$, where obviously $0 \le \xi \le 1$, we can rewrite (2.37) in the form

$$P = \int_0^{\xi_2} \psi(\xi)\,d\xi = \int_0^{\xi_2} 2n^2 \xi e^{-n^2\xi^2}\,d\xi, \tag{2.41}$$

where

$$\xi_2 = \frac{R}{nr_0} = \frac{m}{n}, \quad m \le n.$$

The condition $m \le n$ appears because when $m > n$ the region of existence of the probability density becomes smaller than the size of the detector, and the total probability P is constant and independent of m.

As the thickness of the scatterer is reduced, the quantity $\langle\theta^2\rangle$ decreases and hence r_0 also decreases. For fixed n and R we necessarily have $m > n$, i.e., constant probability. To

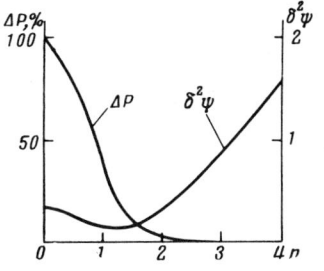

Fig. 9. Dependence of the relative variance and the error in the total probability as functions of n.

analyze the variance we must take the upper limit for m, i.e., m = n or $\xi = 1$:

$$\delta^2\psi = \frac{\frac{n}{2}\sqrt{\frac{\pi}{2}}\,\Phi(2n) - n^2 e^{-2n^2}}{(1 - e^{-n^2})^2} - 1. \qquad (2.42)$$

Figure 9 shows a plot of $\delta^2\psi$ as a function of n, and the error $\Delta P = e^{-n^2}$ in the total probability, which is connected with the restricted range of existence of P.

It is clear from Fig. 9 that there are two conflicting tendencies: with increasing n the error ΔP decreases but the relative variance increases, and vice versa. If we may assume that the error ΔP is of the order of a few percent, n can be taken to be of the order of 2.

The restriction on the range of existence of P introduces additional complications in the determination of the range of integration. While earlier we integrated over the area of the last scatterer or detector, in the present case the range of integration is determined by the intersection of the contour of the scatterer and the ellipse formed by cutting the probability cone of angular aperture $n\sqrt{\langle\theta^2\rangle_i}$ with the plane of the scatterer (Fig. 10). This region is variable and its outline is difficult to establish in general, so that additional machine time is necessary. Moreover, in the Monte Carlo method these regions must be inscribed in each test into unit squares, and an average must be taken of the product of the integrand and the variable areas defined by the corresponding contours S_d.

To simplify the problem we can always integrate over the area of a circle produced by cutting the scattering cone with a plane perpendicular to its axis and passing through the point of intersection of the cone axis and the plane of the last scatterer (detector) (Fig. 11).

Figure 11 shows the section of the scattering cone defined by a plane passing through the cone axis and parallel to the x axis. The trace of this plane is shown by the straight line $(O'_i O_{i+1})$. Moreover, Fig. 11 shows (on the right) the section of the cone by a plane perpendicular to its axis.

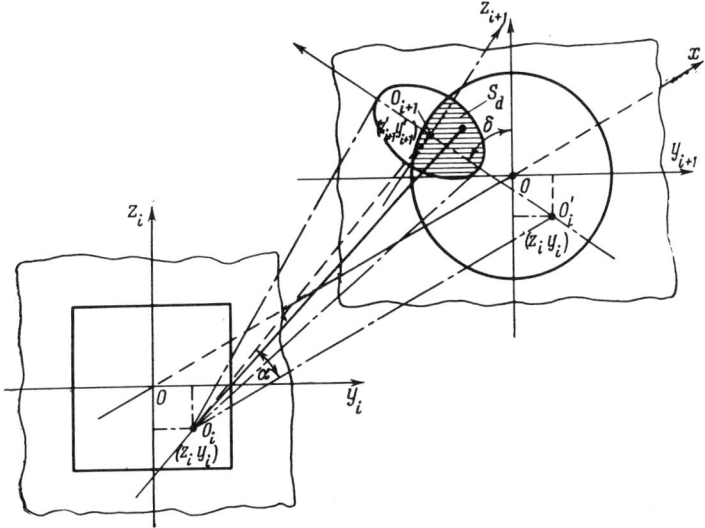

Fig. 10. Definition of the range of integration.

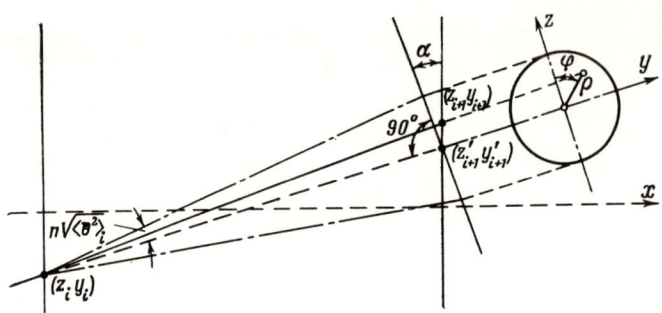

Fig. 11. Section of the scattering cone by the plane $(O_i\ Q'_i\ Q_{i+1})$.

Let us take in this section a coordinate system (zy) so that the z axis lies in the plane $(O, O'_i O_{i+1})$.

We shall specify the position of an arbitrary point in this coordinate system by the two functions ρ and φ, where

$$\rho = \xi_1 \rho_0 = \xi_1 n \sqrt{\langle\theta^2\rangle}\, l \text{ and } \varphi = 2\pi\xi_2,$$

and ξ_1 and ξ_2 are variables, while l is the length of the segment. It is clear that $0 \leq \xi_1 \leq 1$ and $0 \leq \xi_2 \leq 1$. The probability that the scattered particle will pass through the area element $dS = \rho\, d\rho\, d\varphi$ whose center has the coordinate ρ and φ is given by

$$dP = 2n^2 \xi_1 e^{-n^2 \xi_1^2}\, d\xi_1\, d\xi_2. \tag{2.43}$$

The coordinates of this point in the (zy) system are

$$z = \rho \cos\varphi,\ y = \rho \sin\varphi.$$

To transform to the set of coordinates attached to the (i+1)-th scatterer we must rotate the coordinate system (zy) through an angle α in the $[O_i, O'_i, O_{i+1}]$, plane and through an angle δ in the plane of the scatterer, and then perform a parallel translation from the point O_{i+1} to O. We finally obtain

$$y_{i+1} = y'_{i+1} + \frac{\rho}{\cos\alpha - \rho/l\, \sin\alpha\, \cos\varphi}(\cos\alpha\, \cos\delta\, \sin\varphi + \cos\varphi\, \sin\delta), \tag{2.44}$$

$$z_{i+1} = z'_{i+1} + \frac{\rho}{\cos\alpha - \rho/l\, \sin\alpha\, \cos\varphi}(\cos\varphi\, \cos\delta - \sin\varphi\, \sin\delta\, \cos\alpha), \tag{2.45}$$

where

$$\varphi = 2\pi\xi_2 \text{ and } \rho = nl\xi_1 \sqrt{\langle\theta^2\rangle_i}, \tag{2.46}$$

$$l = \sqrt{(L_{i+1} - L_i)^2 + (y'_{i+1} - y'_i)^2 + (z'_{i+1} - z_i)^2}, \tag{2.47}$$

$$\cos\alpha = \frac{L_{i+1} - L_i}{l},\ \sin\alpha = \frac{1}{l}\sqrt{(y'_{i+1} - y_i)^2 + (z'_{i+1} - z_i)^2}, \tag{2.48}$$

$$\cos\delta = \frac{z'_{i+1} - z_i}{\sqrt{(y'_{i+1} - y_i)^2 + (z'_{i+1} - z_i)^2}},\ \sin\delta = \frac{y'_{i+1} - y_i}{\sqrt{(y'_{i-1} - y_i)^2 + (z'_{i+1} - z_i)^2}}. \tag{2.49}$$

In these expressions y_{i+1} and z_{i+1} are the coordinates of the point in the plane of the next scatterer which is reached by the particle as a result of scattering in the previous filter. The coordinates of the point of entry into the scatterer are $(y_i z_i)$, while z'_{i+1}, y'_{i+1} are the coordinates of the point of intersection of the initial direction of the particle with the plane of the next scatterer.

The probability given by (2.43) must be multiplied by a function g_p which is defined by

$$g_p = \begin{cases} 1, \text{ if the point } (L_{i+1}, y_{i+1}, z_{i+1}) \\ \quad \text{falls within the contour of the} \\ \quad \text{next scatterer,} \\ 0, \text{ in the opposite case.} \end{cases} \quad (2.50)$$

If there is a diaphragm between two scatterers we must verify the condition that the particle has passed through this diaphragm (g_{s_i}).

§4. Determination of the Range of Integration for the Entire Integrand

Limits for the Photon Energies. Figure 12 shows the projection onto the xy plane of the geometry of the experiment. The minimum gamma-ray energy will be determined from the kinematics of the problem in terms of the minimum angle θ_{min} and the minimum range $R_1 = \Sigma R_i$, where R_i is the thickness of the individual scatterers, $E_{min} = aR_1^b$ is the minimum proton energy.

$$\theta_{min} = \theta_0 - \theta_1, \quad \theta_1 = \frac{1}{\rho_1^2 + L_1^2}\left[L_1\sqrt{L_1^2 + \rho_1^2 - R^2} - \rho_1 R\right], \quad (2.51)$$

where R is the target radius and ρ_1 is the radius or the half-width of the first scatterer. The minimum gamma-ray energy, for example, in the Compton effect is given by

$$k_1 = \frac{mE_{min}}{p_{min}\cos\theta_{min} - E_{min}}, \quad (2.52)$$

where m and p are the proton mass and momentum.

The first estimate for the maximum gamma-ray energy can be found from R_{max} and θ_{max} using (2.52), where

$$R_{max} \approx \Sigma R_i R_i/\cos\alpha_{i\,max} + R_m^{max} + \Delta R, \quad \theta_{max} \approx \theta_0 + \tan^{-1}\rho_1/L_1.$$

In this expression, α_{max} is the maximum angle of the trajectory between the filters, R_m^{max} is the maximum range in the target, and ΔR is the telescope range interval.

Let us compare the resulting estimate (k'_2) with the maximum energy k_0 in the bremsstrahlung spectrum. If $k'_2 \gtrsim k_0$ then we can take k_0 as the maximum energy k_2. If, on the other hand, $k'_2 < k_0$ then we must calculate the maximum multiple-scattering angle in the target, $\sqrt{\langle\theta^2\rangle_{max}}$, for proton ranges $R_1 + R_m^{max}$ and R_1.

Let us now determine the new maximum angle of escape for the proton

$$\theta'_{max} = \theta_{max} + n\sqrt{\langle\theta^2\rangle_{max}} \quad (2.53)$$

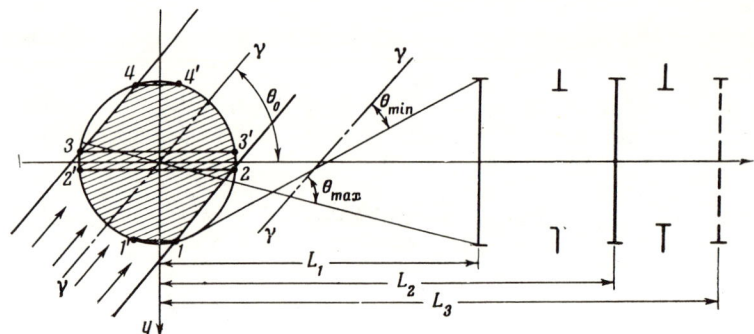

Fig. 12. Determination of the limits of integration.

and the second estimate for the maximum energy from θ'_{max} and R_{max}. We again compare k''_2 with k_0: if $k''_2 \geq k_0$ the maximum energy is $k_2 = k_0$, and if $k''_2 < k_0$ the maximum energy is $k_2 = k''_2$.

Limits for the Variables within the Target. The limits for z are constants. Because of the symmetry of the active volume with respect to the xy plane, we shall consider only one half of this volume, i.e., $0 \leq z \leq r_1$, where r_1 is the beam radius. The limits for y are functions of z. Figure 12 shows the section of the target by a plane parallel to the xoy plane at a height z. In general, $(d = \sqrt{r^2 - z^2} > R \cos \theta_0)$ there are five regions:

$$\begin{aligned}
-R \leq y \leq y_1 \quad & -\sqrt{R^2 - y^2} \leq x \leq \sqrt{R^2 - y^2}, \\
y_1 = & -d \cos \theta_0 - \sin \theta_0 \sqrt{R^2 - d^2}, \\
y_1 \leq y \leq y_2 \quad & -\sqrt{R^2 - y^2} \leq x \leq \frac{1}{\sin \theta_0}(y \cos \theta_0 + d), \\
y_2 = & -d \cos \theta_0 + \sin \theta_0 \sqrt{R^2 - d^2}, \\
y_2 \leq y \leq y_3 \quad & -\sqrt{R^2 - y^2} \leq x \leq \sqrt{R^2 - y^2}, \\
y_3 = & \, d \cos \theta_0 - \sin \theta_0 \sqrt{R^2 - d^2}, \\
y_3 \leq y \leq y_4 \quad & \frac{1}{\sin \theta_0}(y \cos \theta_0 - d) \leq x \leq \sqrt{R^2 - y^2}, \\
y_4 = & \, d \cos \theta_0 + \sin \theta_0 \sqrt{R^2 - d^2}, \\
y_4 \leq y \leq R \quad & -\sqrt{R^2 - y^2} \leq x \leq \sqrt{R^2 - y^2}.
\end{aligned} \quad (2.54)$$

When $d \leq R \cos \theta_0$ there are only three regions:

$$\begin{aligned}
y_1 \leq y \leq y_2 \quad & -\sqrt{R^2 - y^2} \leq x \leq \frac{1}{\sin \theta_0}(y \cos \theta_0 + d), \\
y_2 \leq y \leq y_3 \quad & -\sqrt{R^2 - y^2} \leq x \leq \sqrt{R^2 - y^2}, \\
y_3 \leq y \leq y_4 \quad & \frac{1}{\sin \theta_0}(y \cos \theta_0 - d) \leq x \leq \sqrt{R^2 - y^2}.
\end{aligned} \quad (2.55)$$

If $\theta_0 > 45°$, then $y_2 < y_3$ and the limits are modified:

$$\begin{aligned}
\text{(a)} \quad -R \leq y \leq y_1 \quad & -\sqrt{R^2 - y^2} \leq x \leq \sqrt{R^2 - y^2}, \\
y_1 \leq y \leq y_3 \quad & -\sqrt{R^2 - y^2} \leq x \leq \frac{1}{\sin \theta_0}(y \cos \theta_0 + d),
\end{aligned}$$

$$y_3 \leqslant y \leqslant y_2 \quad \frac{1}{\sin\theta_0}(y\cos\theta_0 - d) \leqslant x \leqslant \frac{1}{\sin\theta_0}(y\cos\theta_0 + d),$$

$$y_2 \leqslant y \leqslant y_4 \quad \frac{1}{\sin\theta_0}(y\cos\theta_0 - d) \leqslant x \leqslant \sqrt{R^2 - y^2},$$

$$y_4 \leqslant y \leqslant R \quad -\sqrt{R^2 - y^2} \leqslant x \leqslant \sqrt{R^2 - y^2},$$

(2.56)

(b) if $d \leq R\cos\theta_0$ there are only the three middle regions in (a).

Boundaries of a Fictitious Area in the Plane of the First Scatterer Which Specifies the Initial Direction of Particles in the Target.
We have restricted the region of existence of the angular distribution function P by the limits of $n\sqrt{\langle\theta^2\rangle}$. Let us therefore take as the linear dimensions of this area the corresponding maximum linear dimension of the first scatterer increased by an amount corresponding to the additional angle $n\sqrt{\langle\theta^2\rangle_{max}}$, where $\langle\theta^2\rangle_{max}$ is the maximum square of the multiple-scattering angle in the target.

Therefore, if ρ_{y1} and ρ_{z1} are the maximum half-widths of the first scatterer, we have

$$\rho'_y = L_1 \tan[n\sqrt{\langle\theta^2\rangle_{max}} + \tan^{-1}\rho_{y1}/L_1], \tag{2.57}$$

$$\rho'_z = L_1 \tan[n\sqrt{\langle\theta^2\rangle_{max}} + \tan^{-1}\rho_{z1}/L_1]. \tag{2.58}$$

To determine $\langle\theta^2\rangle_{max}$ we must calculate the maximum range of the particle from K_2 and θ_{min}, using the kinematics of the reaction under investigation:

$$R'_{max} = f(k_2, \theta_{min}).$$

If $R'_{max} \geq R_m^{max} + R_1$, where R_m^{max} is the maximum thickness of the target and R_1 is the minimum range in the telescope, then $\langle\theta^2\rangle_{max}$ will be calculated from $R_m^{max} + R_1$ and R_1. If $R'_{max} < R_m^{max} + R_1$, then $\langle\theta^2\rangle_{max}$ will be determined from R'_{max} and R_1. If there are kinematic restrictions on the maximum angle of escape of the protons (and this will always occur if $k_2 = k_0$), then the horizontal half-width in the large-angle region will be substantially reduced. The vertical dimensions and the horizontal half-width in the small-angle region will, as before, be determined by the linear dimensions of the first scatterer and by scattering in the target.

We note that in those cases where the limits for any particular variable are difficult to establish exactly, it is usual to assume extremal values. This leads to a loss in the number of successful tests but, at the same time, it ensures that the region of integration will not be accidentally reduced.

Bearing in mind the above limitations on the function P, we may now write the following expression for the integral I:

$$I = \frac{2N_0\sigma}{A} \int_{k_1 0}^{k_2 r_1} \int_{y(z)}^{} \int_{x(y,z)}^{} \int_{-\rho'_y}^{+\rho'_y} \int_{-\rho'_z}^{+\rho'_z} \int_0^1 \int_0^1 \cdots \int_0^1 \int_0^1 f_n(k) \frac{d\Pi}{dS} \frac{\cos\alpha'_1}{l'^2_1} \times$$

$$\times l^{-n_2 \sum_{i=1}^{m} \xi_{1i}^2} \prod_i (2n^2 \xi_{1i} g_{S_i} g_{P_i}) g_R \, dk \, dx \, dy \, dz \, dy'_1 \, dz'_1 \, d\xi_{11} \, d\xi_{21} \ldots d\xi_{1n} \, d\xi_{2m}. \tag{2.59}$$

Continuing now with the Monte Carlo method (method of statistical tests), let us inscribe the volume of the target into the parallelepiped $2R \times 2R \times r_1$. Next, let us change the variables in Eq. (2.22a), and transform to the constant limits 0-1 in all the integrals. We recall, however, that successful tests will be those for which the random variables fall on the region of integration defined by Eqs. (2.54) and (2.55). We then have

$$I = 32 R^2 r_1 \rho'_y \rho'_z \, \Delta k \int_0^1 \ldots \int_0^1 f_n(k) \frac{d\Pi}{dS} \frac{\cos \alpha_1^1}{l_1'^2} e^{-n^2 \sum_{i=1}^m \xi_{1i}^2} \times$$

$$\times \prod_i (2n^2 \xi_{1i} g_{S_i} g_{P_i}) g_R \, d\xi_k \, d\xi_x \ldots d\xi_{z'} \, d\xi_{11} d\xi_{12} \ldots d\xi_{1m} d\xi_{2m}. \tag{2.60}$$

Figure 13 shows the scheme for evaluating the integral. We shall illustrate it by an example in which the telescope consists of two scatterers, three diaphragms, and a detector.

The sequence of operations is as follows:

1. We select 12 random numbers which are uniformly distributed over the segment 0-1 and, for example, we use the first six of these to establish the following physical variables: k, photon energy; x, y, z, coordinates in the targets; and y_1', z_1', coordinates in the plane of the first scatter which specify the direction of the particle. As a preliminary step, we verify whether the coordinates of the point in the target (x and y) satisfy the condition of entry into the active volume, i..e, if $y_l \leq y < y_{l+1}$, $l = 1, 2, 3$, then x enters the admissible interval [see Eq. (2.54)].

2. Next, we determine the angle between the straight line [(xyz), $(L_1 y_1' z_1')$] and the beam axis ($\gamma\gamma$):

$$\cos \theta = \frac{(L_1 - x) \cos \theta_0 + (y_1' - y) \sin \theta_0}{\sqrt{(L_1 - x)^2 + (y_1' - y)^2 + (z_1' - z)^2}}. \tag{2.61}$$

3. The proton energy and its range R_H in hydrogen are determined from the angle θ and the photon energy k, using the kinematics of the problem and the relation $R = aE^b$.

4. We must next calculate the coordinates of the points of intersection of the straight line [(xyz), $(L_1 y_1' z_1')$] with the surface of the target:

$$x_0 = x + \nu(L_1 - x), \quad y_0 = y + \nu(y_1' - y), \quad z_0 = z + \nu(z_1' + z), \tag{2.62}$$

$$\nu = \frac{\sqrt{R^2[(L_1-x)^2 + (y_1^1-y)^2] - (L_1 y - xy_1')^2} - [x(L_1-x) + y(y_1'-y)]}{(L_1 - x)^2 + (y_1' - y)^2}, \tag{2.63}$$

where R is the radius of the target.

5. The next step is to calculate the range in the target:

$$l_0 = \nu l_1' = \nu \sqrt{(L_1 - x)^2 + (y_1' - y)^2 + (z_1' - z)^2}, \quad R_m^H = l_0 \rho_{H_2}. \tag{2.64}$$

6. The residual range $R_{x1}^H = R_H - R_m^H$ is then calculated and converted to the range in aluminum. The following conditions must then be verified:

$$R_1 + R_2 \leq R_{x_1}^{Al} \leq (R_1/\cos \alpha_{1 \max} + R_2/\cos \alpha_{2 \max} + \Delta R/\cos \alpha_{3 \max}). \tag{2.65}$$

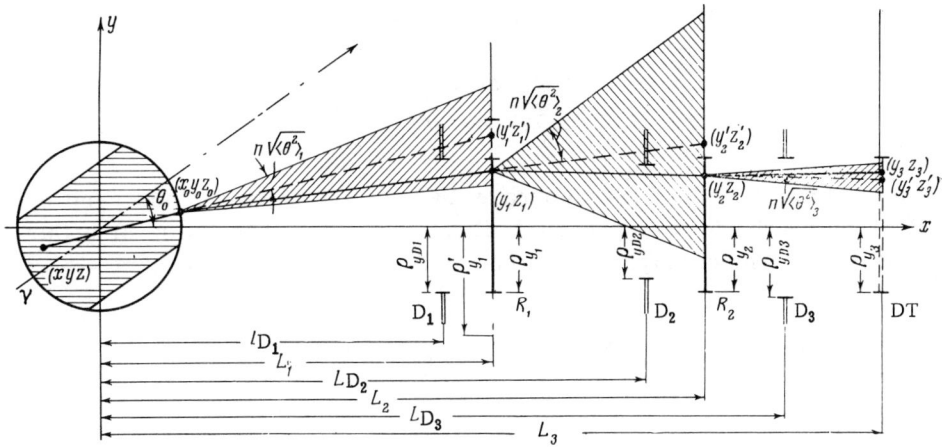

Fig. 13. Scheme for evaluating the integral.

7. If condition (2.65) is satisfied we calculate $\sqrt{\langle\theta^2\rangle_1}$ from Eq. (2.14), using R_H and R_{x1}^H (in hydrogen).

8. Next, we determine the length of the segment $[(x_0 y_0 z_0), (L_1 y_1' z_1')]$, and using $\langle\theta^2\rangle_1$, calculate the maximum radius of scattering in the plane perpendicular to the straight line $[(x_0 y_0 z_0), (L_1 y_1' z_1')]$ and passing through the point $(L_1 y_1' z_1')$:

$$\Delta l_1 = l_1' - l_0, \quad \rho_{01} = n \Delta l_1 \sqrt{\langle\theta^2\rangle_1}. \tag{2.66}$$

9. From the random numbers ξ_7 and ξ_8 we determine the coordinates in the perpendicular plane:

$$\rho_1 = \xi_7 \rho_{01} \quad \text{and} \quad \varphi_1 = 2\pi\xi_8. \tag{2.67}$$

10. The next step is to find the point of intersection of the line passing through the points $(x_0 y_0 z_0)$ and $(\rho_1 \varphi_1)$ and the plane of the first scatterer:

$$y_1 = y_1' + \frac{\rho_1}{\cos\alpha_1' - \rho_1/\Delta l_1 \sin\alpha_1' \cos\varphi_1}(\cos\alpha_1' \cos\delta_1 \sin\varphi_1 + \sin\delta_1 \cos\varphi_1), \tag{2.68}$$

$$z_1 = z_1' + \frac{\rho_1}{\cos\alpha_1' - \rho_1/\Delta l_1 \sin\alpha_1' \cos\varphi_1}(\cos\varphi_1 \cos\delta_1 - \sin\varphi_1 \sin\delta_1 \cos\alpha_1'), \tag{2.69}$$

$$\cos\alpha_1' = \frac{1}{l_1'}(L_1 - x); \quad \sin\alpha_1' = \frac{1}{l_1'}\sqrt{(y_1' - y)^2 + (z_1' - z)^2}, \tag{2.70}$$

$$\cos\delta_1 = \frac{z_1' - z}{\sqrt{(y_1' - y)^2 + (z_1' - z)^2}}; \quad \sin\delta_1 = \frac{y_1' - y_1}{\sqrt{(y_1' - y)^2 + (z_1' - z)^2}} \tag{2.71}$$

and verify the condition that the point $(L_1 y_1 z_1)$ lies in the plane of the first scatterer. If, for example, the contour of the first scatterer is a circle, specified by

$$y^2 + z^2 = r_1^2, \quad x = L_1, \tag{2.72}$$

then the conditions that it will be entered are of the form

$$-r \leqslant y_1 \leqslant r_1,$$
$$-\sqrt{r_1^2 - y_1^2} \leqslant z_1 \leqslant \sqrt{r_1^2 - y_1^2}. \tag{2.73}$$

11. The condition that the first diaphragm has been traversed is now verified. Let the distance to the plane of the first diaphragm be L_{D_1}, and let its shape be rectangular with half-widths $\rho_{D_{y1}}$ and $\rho_{D_{z1}}$, respectively, so that the conditions that the diaphragm will be traversed are

$$-\rho_{Dy1} \leqslant y_{D1} \leqslant \rho_{Dy1},$$
$$-\rho_{Dz1} \leqslant z_{D1} \leqslant \rho_{Dz1}, \tag{2.74}$$

where y_{D1} and z_{D1} are the coordinates of the point of intersection of the straight line $[(x_0 y_0 z_0), (L_1 y_1 z_1)]$ with the plane of the first diaphragm:

$$y_{D1} = y_0 + (y_1 - y_0)\frac{L_{D1} - x_0}{L_1 - x_0},$$
$$z_{D1} = z_0 + (z_1 - z_0)\frac{L_{D1} - x_0}{L_1 - x_0}. \tag{2.75}$$

12. If the conditions given by Eqs. (2.74) and (2.73) are satisfied, we determine the length of the segment $[(x_0 y_0 z_0), (L_1 y_1 z_1)]$, and the cosine of the angle between this line and the x axis:

$$l_1 = \sqrt{(L_1 - x_0)^2 + (y_1 - y_0)^2 + (z_1 - z_0)^2}, \tag{2.76}$$

$$\cos \alpha_1 = \frac{1}{l_1}(L_1 - x_0). \tag{2.77}$$

13. We now calculate the range in the first scatterer and the residual range, and verify the condition that the particle enters the telescope interval:

$$R_{x2}^A = R_{x1}^A - R_1/\cos \alpha_1, \tag{2.78}$$

$$R_2 \leqslant R_{x2}^A \leqslant (R_2/\cos \alpha_{2\,\max} + \Delta R/\cos \alpha_{3\,\max}). \tag{2.79}$$

14. The direction of $[(x_0 y_0 z_0), (L_1 y_1 z_1)]$ is then continued until it cuts the plane of the second filter. The coordinates of the point of intersection and the length of the segment are then calculated:

$$y_2' = y_1 + (y_1 - y_0)\frac{L_2 - L_1}{L_1 - x_0}, \tag{2.80}$$

$$z_2' = z_1 + (z_1 - z_0)\frac{L_2 - L_1}{L_1 - x_0}, \tag{2.81}$$

$$\Delta l_2 = l_1 \frac{L_2 - L_1}{L_1 - x_0}. \tag{2.82}$$

15. The quantity $\langle \theta^2 \rangle_2$ is now determined from R_{x1}^A and R_{x2}^A, and ρ_{02} from

$$\rho_{02} = n\sqrt{\langle \theta^2 \rangle_2} \Delta l_2.$$

16. The random numbers ξ_9 and ξ_{10} are then used to establish the variables ρ_2 and φ_2 [Eq. (2.58)], and the coordinates y_2 and z_2 of the point of intersection of the straight line $[(L_1 y_1 z_1), (\rho_2 \varphi_2)]$ with the plane of the second scatterer [from (2.58)-(2.71)], and the conditions that the particle will enter the second scatterer and pass through the second diaphragm are verified using (2.73) and (2.74).

17. The length of the segment, $[(L_1 y_1 z_1), (L_2 y_2 z_2)]$ and the angle between it and the x axis are determined next:

$$l_2 = \sqrt{(L_2 - L_1)^2 + (y_2 - y_1)^2 + (z_2 - z_1)^2}, \qquad \cos \alpha_2 = 1/l_2 (L_2 - L_1).$$

18. By analogy with step 13 we calculate the residual range and verify the condition that the particle enters the telescope interval:

$$R_{x3}^A = R_{x2}^A - R_2/\cos \alpha_2,$$
$$0 \leqslant R_{x3}^A \leqslant \Delta R/\cos \alpha_{3\,\text{max}}. \tag{2.83}$$

19. By analogy with steps 14, 15, and 16 we then calculate the mean square of the multiple-scattering angle when the protons pass through the second scatterer, and use the random numbers ξ_{11} and ξ_{12} to determine the point of entry in the plane of the detector. The condition of entry into the detector contour and the passage through the diaphragms are then verified.

20. By analogy with step 17 we determine the angle α_3 and verify the last condition

$$R_{x3}^A \leqslant \Delta R/\cos \alpha_3. \tag{2.84}$$

It is assumed that the range interval of the telescope is specified by a certain amount of matter, and may vary slightly, depending on the true trajectory of the proton.

21. If all the preceding conditions have been satisfied, the test is successful. We can then calculate the integrand

$$F_l \equiv f_n(k) \frac{d\Pi}{dS} \frac{\cos \alpha_1'}{l_1'^2} e^{-n^2(\xi_7^2 + \xi_9^2 + \xi_{11}^2)} (2n^2)^3 \xi_7 \xi_9 \xi_{11} \tag{2.85}$$

and introduce it into the summation unit. If any of the above conditions is not satisfied the test is unsuccessful, i.e., $F_l = 0$ and the calculations are repeated again.

22. According to Eq. (2.26), the following quantity is used as an estimate for the integral:

$$I = 32 R^2 r_1 \rho_y' \rho_z' \Delta k \frac{1}{N} \sum_l F_l, \tag{2.86}$$

where the relative variance is

$$\delta^2 I = \frac{\sum F_l^2}{(\sum F_l)^2} - 1/N, \tag{2.87}$$

and N is the total number of tests. If the output B has been measured with a relative error δB, the cross section is given by

$$\bar{\sigma} = \frac{A}{N_0 \rho} \frac{B}{I}, \quad \delta(\bar{\sigma}) = \sqrt{\delta^2 I + \delta^2 B}. \tag{2.88}$$

CHAPTER III

Program for Calculations on the M-20 Computer

§1. Description of Program

The block diagram for the program is shown in Fig. 14. This program has recently grown very considerably by the inclusion of auxiliary elements, and has become quite complicated. Let us consider the individual elements of the block diagram.

The main program (MP) is a realization of the algorithm described in the previous chapter. It can be divided into four sections. The initial section computes the limits for the photon energies and the boundaries of the area specifying the initial direction of the protons. The necessary constants and parameters used later are also calculated at this stage. Sections designated "Kinematics" and "Multiple Scattering" (MS) are used here.

In the "Kinematics" section the energies of the incident gamma rays are calculated from given proton angle and energy, and the proton energies are found from given gamma-ray angle and energy. There are two types of kinematics for the reactions which we are investigating:

$$\gamma + p \rightarrow \gamma' + p,$$
$$\gamma + p \rightarrow \pi^0 + p.$$

The quantity $n\sqrt{\langle\theta^2\rangle}$, where $\langle\theta^2\rangle$ is the mean square multiple-scattering angle, is computed in MS. The variables are the proton range prior to entry into the scatterer and after this entry [Eq. (2.14)]. An address register is used to change the constants describing the material of the scatterer. If AR = 0 then the scattering takes place in hydrogen, and if AR = 0010 the material is aluminum.

The random number loop is then begun. At the beginning of the loop we have the generator of random numbers distributed uniformly in the interval 0-1. The six random numbers for the energy of the gamma ray, the coordinates of the point in the target, and the coordinates in the plane of the first scatterer, which specify the direction, are selected here. This is followed by the analysis of entry into the active volume, the range at exit from the target is cal-

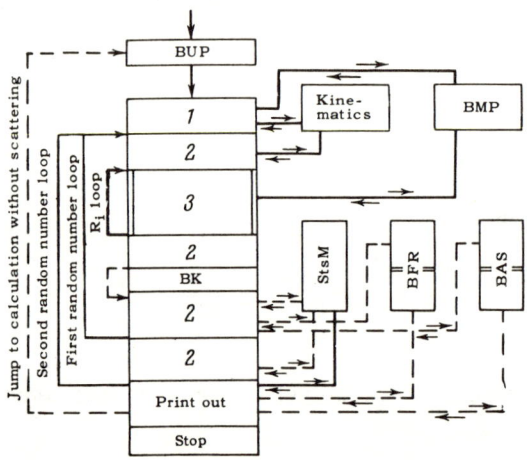

Fig. 14. Block diagram of the program.

culated, and the first very approximate verification of the condition for entry into the range interval of the telescope.

The R_i loop is then begun. The number of scatterers is specified numerically, the maximum number being restricted to five. Scattering in the target is included in the first run through the loop. The random-number generator is located within the loop and produces two random numbers in each run through the loop. At the end of the loop the integrand is calculated and the sums are stored. In addition to the sums ΣF and ΣF^2 the following sums are also accumulated $\Sigma \theta F$, $\Delta \theta^2 F$, $\Sigma k F$, and $\Sigma k^2 F$. They are used to obtain the angular and energy-distribution parameters. The number of successful tests is counted at this stage. The total number of tests is stored for each return to the random-number generator which lies at the beginning of the external loop. The random-number loop terminates in two stages. Firstly, the loop ends when the number of successful tests is a multiple of a given number, for example, $n_0 = 100$. At this stage the accumulated sums are printed out together with the value of the integral and its relative variance. If necessary, the center-of-mass results can be printed out together with the angular and energy distribution parameters.

The relative accuracy of the integral is then compared with the specified figure. The specified accuracy can either be constant (usually taken to be one-third of the accuracy of the output) or variable and specified from machine control panel.

The random-number loop terminates when the specified accuracy is achieved. The results are then printed out in the laboratory system and in the center-of-mass system. The main results are also recorded on punched tape. The program-control section then takes over and a comparative calculation without scattering is performed.

The last stage usually occupies very little time and has practically no effect on the overall time of computing. Moreover, the main program includes a correlation section which lies within the random-number loop and analyzes the conditions for the recording of the second particle.

Thus, when the reaction $\gamma + p \rightarrow \gamma'$ is being analyzed the angular and energy parameters of the protons and the energy of the primary gamma ray are used to determine the coordinates of the scattered gamma ray in the plane of the gamma-ray counter. In this case, successful tests will only be those for which the scattered gamma ray enters the aperture of the gamma-ray counter (recording condition). When this section is operating the additional sums $\Sigma \theta_\gamma F$ and $\Sigma \theta_\gamma^2 F$ are stored and then used to determine the mean angle of escape and its variance for the scattered gamma rays.

This section is usually by-passed and is brought into operation by means of a special conditional routine available at the control panel.

Distribution-Function Section (DF). Before we describe this program section we must consider the physical significance of the integral I. This integral represents a certain volume ΔV in the phase space in which the reaction under investigation takes place. Its limits are determined by the energy interval of the telescope and the geometry of both telescope and target. The total volume V in which the given reaction is allowed is determined by the entire energy spectrum of the primary gamma rays, and the presence of possible kinematic and energy restrictions unconnected with the apparatus. The magnitude of I is therefore proportional to the probability of detection of the secondary proton. In view of this probabilistic interpretation of I we must regard the integrand F as a probability distribution density over a phase area coinciding with the given region of integration. Therefore, if we desire to consider the distribution of some physical quantity which depends on the same variables we must ascribe to this quantity a weight equal to F in each successful test. This conclusion was used when we were discussing the calculation of the energy and angular distribution parameters for the experimental arrangement. For example, if we wish to determine the average energy of the incident gamma rays and their variance then, in accordance with the foregoing conclusion, we may write

$$\bar{k} = \frac{\int_\Omega k F(x)\, dx}{\int_\Omega F(x)\, dx} \approx \frac{\sum k_l F_l}{\sum F_l},$$

$$D_k = \bar{k}_2 - \bar{k}^2 \approx \frac{\sum k_l^2 F_l}{\sum F_l} - \left(\frac{\sum k_l F_l}{\sum F_l}\right)^2, \tag{3.1}$$

where, for simplicity, x represents the entire set of variables, and Ω denotes the region of integration.

The calculation of such parameters will not, therefore, present particular difficulties, especially if the corresponding sums are stored. However, in some cases, it is extremely interesting to obtain the form of the distribution. For example, in design calculations for an experiment it is possible to deduce the energy spectra of protons in different counters of the telescope, and the spatial distributions of protons within the planes of the individual diaphragms or scatterers. Moreover, it is possible to calculate the spatial and energy distributions for the second particle, so that the geometry, angle of acceptance, and other parameters of the various counters can be suitably chosen.

This problem, namely, the sampling of a given quantity, its storage, and then the synthesis of the distribution function, is tackled by the distribution-function section (DF). It consists of two parts, namely, storage and processing. The store is interrogated after each successful test. This part of the program accumulates the quantities under analysis and the corresponding weights (F_l) in a small section of the operational store. After this part of the store has been filled, all the numbers are transferred to a drum. At the end of the calculation, i.e., when the end of the random-number loop has been reached, the results are processed, i.e., the average $\bar{x} = \Sigma x_l F_l / \Sigma F_l$ and the variance $Dx = \frac{\sum x_l^2 F_l}{\sum F_l} - \left(\frac{\sum x_l F_l}{\sum F_l}\right)^2$ are evaluated. A conversion is then made to the random quantities $y_l = x_l - \bar{x}$. The domain of y_l is divided into intervals of width σ_x/ν, where $\sigma_x = \sqrt{Dx}$, and ν is an integer ≥ 1, which is preset from the control panel prior to the calculation. All the events y_l are then sorted out into the intervals, and the sum of the corresponding weights for each interval is stored. This ends the analysis. A table is then printed out in accordance with a special standard program, showing the frequency and the limit of the interval. For comparison, a table is printed out with weights equal to unity.

If the dum is filled more rapidly than the rate at which the given accuracy is achieved, the program jumps to processing after the last possible protion has been stored. The distribution function section is then bypassed, and the evaluation of the integral continues. The DF program section is constructed so that the analysis can be carried out simultaneously for three different variables. The choice of any particular variable is achieved by a set of corresponding numbers identifying store registers.

Analysis-of-Average Section (AV). It was noted earlier that the estimates of the magnitude of the integral for a sufficiently large number of successful tests are distributed approximately normally. The AV section of the program can be used to analyze the distribution function for the estimates of the integral which are obtained by successive evaluation of the mean values of the function F for a given number of successful tests. Let us suppose that altogether N tests were carried out of which n were successful. Let us take a certain number Δn of successful tests for which we shall evaluate the integral

$$I_j = \frac{1}{N_j} \sum_{l=1}^{\Delta n} F_{l_j},$$

where j is the running number of the estimate, and N_j is the total number of tests corresponding to Δn:

$$j = 1, 2, \ldots, m; \quad \sum_j N_j = N, \quad m = \frac{n}{\Delta n} \; (n - \text{small } \Delta n).$$

We shall take the reciprocal of the variance,

$$W_j = \frac{1}{D_j} = \frac{N_j}{D},$$

as the weight for I_j, where D is the variance of the integrand. The mean and the variance for the entire series will be written in the form

$$I = \frac{\sum_j W_j I_j}{\sum_j W_j} = \frac{\sum_j \sum_l F_{lj}}{\sum_j N_j} = \frac{\sum_{k=1}^N F_k}{N},$$

$$D_I = \frac{1}{(\sum W_j)^2} \sum_j W_j^2 D_j = \frac{D}{N},$$

which is the same as the definition of the average of I and its variance over the entire set of n successful tests.

Now consider the random quantities $y_j = \frac{I_j - I}{\sqrt{D_I}}$ and divide the domain of y_j into intervals of width $\Delta = 1/2 \sqrt{D_I}$. Let us next sum in each interval the weights of the events and determine the frequencies. Assuming a normal distribution law $N(y, 0, 1)$, we shall calculate for these intervals the corresponding probabilities, and compare the resulting empirical distribution with the normal distribution by the χ^2 test [24]:

$$\chi^2 = \sum_{s=1}^{l} \frac{(N_s - Np_s)^2}{Np_s},$$

where l is the number of intervals in which $N_s \geq 10$ (if $N_s < 10$ the particular interval is combined with the next), p_s is the corresponding probability which follows from the normal distribution, N_s is the sum of weights ($N_s = \Sigma N_j$) for events in the given interval, and N is the total number of tests. The number of degrees of freedom will be $l - 3$ since we have determined the estimated average and variance from the same sample.

Apart from estimating the integral, the AV section analyzes four further variables, namely, the distribution of estimates of the range of integration $\Delta n / N_j$, the mean energy of the γ-rays, the mean angle of escape of the proton, and the mean angle of escape of the scattered γ-ray. The quantity Δn is preset during the calculation (from the control panel). Both

DF and AV sections consist of two parts, namely, storage and processing. After Δn successful tests have been reached during the calculation, the program jumps to AV where the average values and their weights are calculated and recorded on the drum. The computation then returns to the main program and the Monte Carlo routine is continued. At the end of the calculation the program again goes over to AV, where the entire stored material is used to construct the distribution functions. The distribution functions are then printed out together with the corresponding parameters, namely, χ^2, the number of degrees of freedom, and the empirical frequency in the range $\pm\sqrt{D_I}$. The program section terminates at this point.

Program Control Section (PC). This section is used to determine the mode of operation of the program and its adjustment. Data on the given mode of operation reaches PC from the numerical data or (largely) from the control panel. Let us briefly consider the individual variants of the calculations.

1. Choice of Process. Information about the particular process (Compton effect or photoproduction of neutral mesons) is contained in the numerical data. In accordance with this information the program control section brings in the necessary kinematics, the program for the conversion to the center-of-mass system, and corrects the correlation section (CB).

2. Main Variant of Operation. Here, the PC adjusts the program to calculation with scattering, specifies the length of the scattering loop and the accuracy of the calculation ($\delta_I = \frac{1}{3}\delta B$). At the end it adjusts the program to calculation without scattering, and transfers the control back to the beginning of the program, so that a comparative calculation can be carried out under these conditions. Moreover, if some additional sections are used in the initial calculations, they are all bypassed at this stage.

3. Calculation without Scattering. The program starts without the inclusion of scattering.

4. Calculation without Scattering in Target. The loop which takes into account scattering in the target is bypassed.

5. pγ Correlations. Here, the correlation section is included.

6. Construction of the Distribution Functions. This is done by the DF section.

7. Analysis of the Distribution of the Averages. This is performed by the AV section.

8. Continued Computation. If, for some reason, the calculation is stopped, a special punched card is introduced to continue the calculation together with the numerical data. The PC introduces the punched card, addresses the accumulated sums to the appropriate registers, and passes the last random number to the random-number unit. The calculation can then be continued. The punched cards carry no information about the DF and AV sections.

9. Effect of "Stretching." Information about the presence of a primary-electron energy distribution is introduced into the standard program for calculating the normalized number of photons. Some of the distribution parameters are calculated and stored.

10. Effect of the Bremsstrahlung Flux-Density Distribution. Here, the parameters of the function $d\Pi/dS$ are computed using the corresponding program section. This part of the program can be bypassed if this function turns out to be practically constant.

11. Modifications in the CM Block. When the distribution function is being computed it is occasionally necessary to analyze certain other parameters in the center-of-mass system, for example, the conversion coefficient, the CM angle, etc. In this case, the CM section is used after each successful test. In this mode the CM section is included in the random-number loop but not entirely. Only that part of it is used in which the main physical quantities are computed.

TABLE 3

r'_0	$\rho'_0 = 0.2$	$\rho'_0 = 0.3$	$\rho'_0 = 0.4$	$\rho'_0 = 0.5$
0.6	0.932 ± 0.009 0.928	0.920 ± 0.009 0.915	0.906 ± 0.0086 0.897	0.860 ± 0.0086 0.874
0.7	0.858 ± 0.0079 0.859	0.846 ± 0.082 0.846	0.826 ± 0.0082 0.827	0.869 ± 0.0076 0.803

In addition to the operations performed by the program control section there are a number of auxiliary operations performed by the program, namely, the printing out of intermediate results in the center-of-mass system, the modification of the required accuracy, and the production of punched cards for intermediate outputs from the random-number loop.

The information for these modes is also taken from the control panel of the M-20 machine and can be prescribed during the computation.

§2. Program Checks

After the program was developed it was used to solve a number of control problems as a check on the individual elements of the program.

1. Computation of the active volume of the target. For this problem we chose the conditions under which the final result could be calculated analytically. The control computation agreed with the analytic calculations to within 0.3% (accuracy of computation 0.88%). The logic of this computation was also verified.

2. Check of that part of the program which gives the probability of detection after passage through a scatterer. The fraction of particles reaching a circular detector of radius R and located at distance l from the scatterer was calculated in [1]. It was assumed there that the primary-particle beam was parallel and uniform within a circle of radius ρ_0. The scattering process was characterized by a parameter r_0 in the plane of the detector. A calculation was then made of the fraction F of particles in the primary beam which reached the detector after passing through the scatterer. We have made a comparison with the results of [1] by modifying the program as follows: 1) a parallel and uniform particle beam of circular cross section having radius ρ_0 was isolated from the target; 2) detection of the particles meant incidence on a circular detector of radius R; 3) the multiple-scattering angle was not calculated, and a specification was simply made of the parameter r_0 in the plane of the detector.

The calculation was carried out with a preset accuracy of 1%. Within this accuracy the agreement with [1] turned out to be very good for different values of $\rho'_0 = \rho_0/R$ and $r'_0 = r_0/R$. This is partly illustrated by the data given in Table 3. In this table the upper row of numbers shows the results taken from the graph in [1].

3. As indicated in the Introduction, we analyzed the Compton effect in 1960 using a somewhat simplified scheme and a manual method of calculation. Scattering in the target was not taken into account and the effect of multiple scattering in the absorbers was allowed for approximately using the results given in [1].

These calculations were repeated on the electronic computer using the above program. Results of computer calculations carried out for the case without scattering agreed to within the accuracy of the original calculation (about 3%) for a number of control points. When scattering was introduced (in the last calculation we took scattering in the target into account) the discrepancy in the entire range of angles varied between 7 and ~40%, tending to reduce the integral. At the most reliable points (small angles) the discrepancy amounted to an average of about 15%. The estimated effect of scattering in the target for this position was 11%. The remaining discrepancy of about 5% could be explained by the more correct method of allowing for multiple scattering in the computer program as compared with the manual calculation.

Conclusions

This work is a logical continuation of the data-processing work which has been continuing over a number of years in our group. Although these problems arise sporadically and, in a way, are subsidiary to the main line of research, they must be considered at each stage before an analysis of the physical experiment is carried out. In this paper we have described the first attempt to transfer most of the numerical work to the computer. The algorithm for the problem and its realization in the program are therefore by no means perfect.

In conclusion, I should like to thank L. N. Shtarkov, G. A. Sokol', and P. S. Baranov for many useful discussions of both general aspects of the problem and specific questions connected with the realization of the algorithm on electronic computers. I am also grateful to E. M. Leikin, V. F. Grushin, and S. P. Denisov of the Lebedev Institute, who read the manuscript and provided valuable suggestions. Finally, I am indebted to the staff of the computational department for their assistance in the calculations.

LITERATURE CITED

1. R. M. Sternheimer, Rev. Sci. Inst., 25:1070 (1954).
2. Atkinson and Willis, High Energy Particle Data, UCRL-2426 (1957).
3. R. M. Sternheimer, Phys. Rev., 118:1045 (1960); 103:S11 (1956).
4. B. Rossi, High-Energy Particles [Russian translation], Gostekhizdat, Moscow (1955).
5. G. Molier, Z. Naturforsch., 3a:78 (1948).
6. G. Molier, Z. Naturforsch., 2a:133 (1947).
7. H. A. Bethe, Phys. Rev., 89:1256 (1953).
8. E. J. Williams, Proc. Roy. Soc. (London), A169:531 (1939).
9. S. A. Goudsmit and J. L. Saunderson, Phys. Rev., 57:24 (1960); 58:36 (1960).
10. H. Snyder and W. T. Scott, Phys. Rev., 76:220 (1949).
11. B. P. Nigam, M. K. Sundaresan, and T. Y. Wu, Phys. Rev., 115:491 (1959).
12. R. H. Dalitz, Proc. Roy. Soc. (London), A206:509 (1951).
13. R. Hofstadter, Rev. Mod. Phys., 28:214 (1956).
14. B. Rossi and K. Greizen, Interaction of Cosmic Rays with Matter [Russian translation], IL, Moscow (1948).
15. E. Jahnke and F. Emde, Tables of Functions [Russian translation], Gostekhizdat, Moscow (1949).
16. L. Eyges, Phys. Rev., 74:1534 (1948).
17. V. F. Grushin and E. M. Leikin, Pribory i Tekh. Eksp., No. 1, p. 52 (1965).
18. L. J. Schiff, Phys. Rev., 83:252 (1951).
19. A. S. Penfold and J. E. Leiss, Analysis of Photo Cross Sections, Preprint, University of Illinois (1958).
20. A. N. Gorbunov, Private Communication.
21. N. P. Buslenko et al., The Monte-Carlo Method [in Russian], Fizmatgiz, Moscow (1962).
22. N. P. Buslenko and Yu. A. Shreider, The Monte-Carlo Method [in Russian], Fizmatgiz, Moscow (1961).
23. V. N. Demidovich and I. A. Maran, Fundamentals of Computational Mathematics [in Russian], Fizmatgiz, Moscow (1963).
24. I. V. Dunin-Barkovskii and N. V. Smirnov, Short Course of Mathematical Statistics for Technological Applications [in Russian], Fizmatgiz, Moscow (1959).

TYPICAL AND ATYPICAL FAILURES OF THE GENERAL-PURPOSE COMPUTER M-20, METHODS OF LOCALIZATION AND ELIMINATION

V. V. Gavrilov, B. G. Minaev, and Yu. V. Stupin

Introduction

The general-purpose electronic computer M-20 contains hundreds of thousands of electronic components, each of which has a limited lifetime. Failure of a single element leads to machine errors, for whose localization and elimination a definite time is required. The development of practical methods for the localization and elimination of failures take on great importance in connection with these questions.

The present article consists of five chapters. Chapter I is devoted to methods of testing and localizing failures from the control console in the step-by-step mode of instruction execution. In Chapter II questions of detecting failures by technical means and various auxiliary circuits are considered. In Chapter III examples of typical and in Chapter IV atypical machine failures and methods for their localization and elimination are presented. In Chapter V circuits are described which permit the operating reliability of M-20 to be increased and the execution of certain operations to be accelerated. Methods for increasing the reliability of the machine are indicated.

The authors express their appreciation to senior scientific collaborator, Candidate of Technical Sciences A. V. Kutsenko for valuable comments and constant support in the work, and the referee of the manuscript, the chairman of the technical council of the KEVM N. M. Yakovlev, who contributed a number of useful suggestions.

CHAPTER I

Failure Detection from the Control Console Logical Analysis

§1. Tests and Test Problems

Testing serves to establish the operability of all devices, blocks, and registers of the machine by means of special test programs. Each single test checks an individual device or

---

* Acknowledgements are due to Mr. Wade B. Holland of the Rand Corporation Computer Sciences Department, who assisted in the decoding of the numerous acronyms and abbreviations appearing in this article. A glossary of these is given in the Appendix.

block of the machine with minimal loading of the other devices and blocks (autonomous test). The number of individual tests depends on the number of machine devices. Tests exist for checking the control, arithmetic, core storage, magnetic [drum] store, and the input − output devices. For each of these several types of tests are used (multiplication, MA to a variable code, etc., for testing the arithmetic unit, test of core storage by a variable code and a repetitive test, etc.). The autonomy of the tests is not absolute; a test for checking one device often eliminates failures in another. This is explained by the fact that the machine as a whole is a system of logically and electrically interconnected devices. To a certain degree, single tests satisfy the requirements of convenient localization and elimination of obvious failures, but are insufficient to detect random intermittent faults.

Example. On stops of the core memory test the 36th bit sometimes is missing in register R1, sometimes in register R2. Change in the nominal − 40 V over the working limits does not influence the character of the fault.

The cause of the fault may be any element in the write − read networks of core. But since the character of the fault is independent of variations in − 40 V, it appears that a RF cell to which this bias is not applied is at fault.

Example. At bias − 44 V an address failure occurs in core. The address is formed in AA correctly. In the "address code" combination at the time the test halts the 12th, 24th, and 36th bits in register R2 are missing, compared to R1.

The core address decoder position consists of elements G1 and RF. Since the failure depends on the bias − 40 V, it is most probable that the element G1 at the input to the 12th bit is being upset.

Example. The contents of register 0010 in the core test are lost at bias − 43 V. Halt on writing showed that sometimes in place of writing in register 0030 register 0010 was being written in.

By means of the oscilloscope it was found that the output signal of element A5 in the 5th position of AA is weaker than in the other positions.

Example. With change of voltage + 300 to + 280 V test halt occurs, where only zeros sometimes appear in register R1, sometimes in register R2.

Hypothesis: writing in core does not always take place over P17 of CSC. Oscilloscope investigation showed that the P17 CSC pulse "floated" on the bus.

These examples show that in the analysis of failure causes it is necessary to consider the machine as a whole and the interaction of its devices with each other.

The repetitive test of core serves to test the elements F_z and partially F_x and CS2. All registers in core are tested repeatedly. At test halts statistical data are collected on the repeatability of the numbers F_z (from the contents of RA) and the bits which have dropped out (from the contents of R1 and R2). The registers are then examined with the oscilloscope and, when the examination does not indicate that the element is defective, elements are interchanged to localize the fault.

Example. With reduction of voltage + 200 to + 185 V the test halted at values of RA equal to 2015, 2057, 2063.

Since $F_{zy} = 20$ is repeated, it is exchanged with $F_{zy} = 21$. After substitution, 21 lights up in the higher-order bits of RA at test halts.

If the repetitive test of core stops when the nominal voltages are varied within given limits at a constant value of RA and exchange of elements does not help, this indicates a defect in the corresponding digit line and the need to go around it.

TYPICAL AND ATYPICAL FAILURES OF THE M-20

When testing arithmetic and control units a combined test is used which consists of several individual tests: "0 absent," ω, AA, control, multiplication, overall, MA, and EA in a variable code, and a special control program "header." Each test verifies certain blocks of these devices. The tests contain several examples which are executed one after the other. When halt occurs at an error in the computation of a given example, it is possible to exit to this example by transferring the machine from "automatic" mode to "single-step" mode and pressing the "start CSC" button several times (depending on the example). In the AA and Control tests intermittent faults are usually very rare, occurring most frequently in the tests "0 absent," ω, multiplication, and overall. The tests MA and EA in variable code are inconvenient for analysis and the search for causes of faults, and may be used as checks. The test "0 absent" and ω test the operation of the analysis networks. The causes of intermittent faults of the machine in halts of these tests are unreliable operation of the amplifiers A_2, operating from assemblies of 12 positions of R_1, MA (particularly on the side of oscillation) and more rarely, faults in the operation of individual flipflops for analysis of "0" R1, "0" Ad, "0" R2. The multiplication and overall tests basically check the operation of the large-scale arithmetic unit networks (elements AM, CM, DF, A2). Intermittent failures in the local control circuits of the operations and the arithmetic synchronization block are substantially rarer. The search for failures by these tests requires a careful analysis of the character of the failure, since the causes of failures may be quite varied. We present here several examples.

Example. In the ω test halts occur on the execution of example 0.14.0100.A2.1000 for <A2> = 1.20.0001.0000.0000. The ω tag remains equal to 1.

Since no failures occur in the other examples, the general part of the R1 analysis circuit (beginning from the gates of A2 and terminating with the input gates to F "ω") operate correctly. It is therefore possible to consider that the error occurred because of the 25th position of R1.

Example. In the test "0 absent" halts occur on execution of the example 0.01.0123.1000.1001. As a result, in place of 1.63.7777.0000.4000 the result 1.63.7777.0000.0000 is obtained, i.e., roundoff has not occurred.

Roundoff in addition depends on the analysis of the contents of register R2, which in the example are not equal to 0. Oscilloscope examination showed that F "0" of R2 is being upset.

Example. In the overall test halts occur at the example 6.44.0001.0002.1000. In place of the correct result 5.13.6314.6314.6315. 5.13.0000.0000.6315 is sometimes obtained.

After oscilloscope examination it was found that the flipflop in the LCOKK circuit which gives the signal ObrMA is being upset.

Tests AU-1 and AU-2 serve for a more careful check of the arithmetic unit. The test AU-1 is rarely applied; usually test AU-2 is used, which consists of a large number of examples checking the arithmetic unit in the execution of various operations and with various combinations of codes. Each example is executed 777_8 times, and then the correctness of the results is checked. Most often, on exit to an example after halt it is possible to obtain a clear fault by varying the tensions -40, -22 or $+200$ V, whose cause may be found by oscilloscope. As in other tests, the cause of halt need not be an intermittent failure in AU, but in other units.

Example. At a bias -44.5 V halt occurs in test AU-2 for RA-0041.

Has storage taken place at address 4641 in the execution of the current example? To establish this we call into IR the instruction of the current example, we block CRA, transfer the machine mode to "single-cycle" and check instruction selection in IR in the mode "erase IR" – "start CSC." We observe that the 6th position is not always selected in IR. Interchanging the

DF elements of the 6th and 1st positions of RK, we find that afterwards the failure has shifted to the 1st position.

Example. In the execution of examples errors occur in many positions. In cycling the examples are correctly executed.

Since the results of all examples in the test are written in addresses 4601 – 4677 of core, it may be assumed that the element $F_{zy} = 46$ is being upset.

Test MM serves to test the operation of the drums and tapes in all modes. When necessary, it may be used to test the output devices. The test is excellent for the write – read circuits of the magnetic memory devices, the quality of the heads for various cases of writing (0 on 1, 1 on 0, 0 on 0, and 1 on 1) and the LCOMM circuit. It is not difficult to find the causes of failures in magnetic storage. Aside from this, failures in writing or reading in magnetic storage are not so dangerous in the solution of problems, as a result of the automatic checking of these circuits. However, certain complications are possible here as well, provoked by intermittent failures in other devices. In this case the natural method is to examine the CCOMM circuits by oscilloscope. Let us illustrate this by examples.

Example. On test halt at − 37 V the contents of registers R1 and R2 disagree in the highest order octet. The channel numbers differ.

An examination of the circuit SPD did not reveal the cause of failure. Further oscilloscopic examination showed that amplifier A2 is being upset in the circuit for generating the signal SdLR41 in the control unit.

Example. On test halt the contents of R1 and R2 disagree strongly. No regularities are observed at the console.

Oscilloscope examination showed that pulse P7 is small in amplitude.

The output test checks printing and punching with various codes and differing volumes of output. Correctness of printing is checked by a visual examination, and punching by fictitious input through the CR of the machine. The test is excellent for checking the operation of the writing and reading circuits, the print number register and the LCOPr circuit. On failures it is fairly difficult to use the oscilloscope, due to the complexity of synchronizing low-frequency intermittent failures. Therefore, elements are frequently interchanged in order to localize the failure. As examples we present the following common cases.

Example. In the 5th position a number is sometimes printed which is less by 2 than the prescribed one, where this failure does not depend on the bias − 40 V. Printing from the simulator is correct.

The element TG2 of the print number register does not depend on the bias − 40 V. After interchange of the 5th and 1st TG2 the failure changed place.

Example. In even positions numbers less by 2 than prescribed are printed. The failures are independent of the voltage − 40 V.

Since the failures occur only in even positions and are the same in character, the cause may be the write – read network in the sixth buffer line.

A printing test is also used to check output, in which the program provides output of only one kind: decimal lines, and separated from them "oblique text."

As a result of their autonomy, the individual tests do not give a complete picture of the operation of the devices as parts of systems, and are rather difficult to manipulate. Therefore, the so-called complex tests have appeared. Using them the machine is tested in a mode approaching that in the solution of problems. The systems CT-1, CT-2, and CT-3 have been developed. Systems CT-1 and CT-2 contain a small number of tests and have not therefore obtained wide circulation. At the present time almost all M-20 machines use the test system CT-3, which is more highly developed than the systems CT-1, CT-2. System CT-3 consists of several individual tests: "0 absent," ω, AA-1, control, multiplication, overall, AU-1, EA cyclic operation, AU-2, repetition of core on constant code, TCCO MM, repetition of core on variable code, MM, MD-1, AA-2, output, random, and the special control program "header" which sets up the order of execution and number of repetitions of each test. "Header" also includes a power-supply test, which is executed before each test to establish the load on the basic voltage -40 V. The test system CT-3 is stored on tape, and before use is transferred to MD-3, on which its operation is based. The system CT-3 has two possible operating modes: individual, in which only one selected test is executed, and cyclic, when the selected tests are executed one after the other, in order. The operating mode of CT-3 and test selection is carried out from NVM-4. The values of the positions of NVM-4 are given in Table 1. In cyclic mode tests No. 1-12, 14, 16, and 20 can operate. Tests No. 13, 15, 17, 21, and 22 may operate only in individual mode.

After input of the auxiliary program "call header of CT-3 from MD-3" the header is transferred from drum to registers 7501-7773 in core. The same program transfers control to register 7666 of header. MD-2 is allocated in header in address code for subsequent testing (the test of MD-2 in header may be avoided by pushing in the 43rd position of NVM-4), and then control is forced to the Aut. STOP test in individual mode. This is necessary in order to test the drum and Aut. STOP circuits before running through the system CT-3. After the Aut. STOP test has been run through once, header analyzes CT-3 mode (1 in position 42 on NVM-4 gives individual mode, 0 gives cyclic mode). If the individual mode of CT-3 is prescribed, then header, determining the number of test selected (positions 24 to 41 of NVM-4) calls it into core (beginning at address 0001), executes power-supply test (the number of times defined by positions 13 to 23 of NVM-4) and transfers control to the selected test. Each test operates a certain number of times. After execution of the test control is transferred to address 7666 in header. The process is then repeated, beginning with the power-supply test. If the cyclic mode is assigned, then CT-3 operates as above, but after the test is terminated and control is transferred to address 7666 of header, the number of the next test selected is determined. This test is then called, the power-supply test is executed, the called test, and again control is transferred to header. This order of operations continues until all selected tests have been executed. Then the process repeats. To monitor the course of the tests core address 7522 is set up on the control console keyboard and the toggle "call codes" is set. Then in the second address of the result register the number of the current test lights up. In individual mode of CT-3 each individual test may be controlled from NVM-4 and positions 1 to 4 of NVM-3. Below we give a brief description of the test included in CT-3 and the control of their operating modes.

The test "0 absent" checks for presence of 0 R1, R2, MA, EA, and AA. The analysis circuits determine generation of the tag ω (0 or 1) by which they are controlled. The test consists of several examples which occur one after the other and are executed in succession. Each example consists of five instructions: a) "0" or "PB" in the power-supply test, b) the example instruction, c) an instruction for testing the correctness of the result according to the tag ω ("36" or "76"), d) stops ("17," "77," or "35"), e) an instruction to return to the start of the example. In the case of erroneous execution of the example halt occurs at one of the instructions listed in (d), with output to R1 of the result of the operation and to R2 the instructions carrying out the example. In the case of "0 absent" R2 halt occurs on error in address 0056 with output of the reference to R1 and the result to R2. On error in examples in "0 absent" AA halt occurs on the instruction 0.77.7777.7777.0000 (selection of instruction "from under itself"). In individual mode 1 in position 1 NVM-3 provides execution of power-supply

TABLE 1. Diagram of Test System Control from NVM-4 in M-20

Positions									
45	44	43	42	41	40	39	38	37	
Test MD from NVM in test 17; Truncation of test 12 Operation with MM in test 16	Test 17 "0 on 0," "1 on 1" Blocking exit to header on cutoff in test 12 Test 13: 0 – complementary code, 1 – direct code Variable zone number in test 16	Run over MD in header Cycling in MM Cycling in defective part in test 17	System mode: "0 – cyclic, 1 – individual	Aut. STOP	Output	AA-2	MD-1	MM	

Positions												
36	35	34	33	32	31	30	29	28	27	26	25	
Repetition of CM=2	CCO MM	Repetition of CM-1	AU-2	Cyclic operations	EA	AU-2	Overall	Multiplication	Control	AA-1	ω	
15	14	13	12	11	10	7	6	5	4	3	2	

Positions											
G_{12} P_3 test ARep	G_{11} Pr_8 move 0 and 1	G_{10} P_2 move 0 and 1	G_9 Pr_{10} minus along diagonal	G_8 P_1 punching proper	G_7 Pr_{10} ladder	G_6 Pr_{10}	G_5 Pr_{10} rows	G_4 Pr_{10} 100 zeros	G_3 Pr_8 ladder	G_2 Pr_8 sevens	G_1 Pr_8 rows

Positions						
24	23	22	21	20	19	
"0 absent"						The 19th to 24th positions give the number of times the direct code is repeated in test 17
						the 13th to 18th positions give the number of times the complementary code is repeated in test 17

Positions												
12	11	10	9	8	7	6	5	4	3	2	1	
	Operation over the entire memory in test 12									Disconnect MD in test 14	Operation of system with MT

<u>Notes</u>: Positions 36 to 25 – CN instructions "50" in test 16; initial RA cutoff in test 12. Positions 24 to 13 – initial address of MM in test 16; number of repetitions of power-supply test, number of repetitions in tests 13 and 15. Positions 12 to 1 – number of codes for tests 14 and 16, final RA cutoff in test 12.

test before one of the examples. Here it is necessary to set up on NVM-2 the code 0.56.0000. $Y+_1$, where Y is the number of one of the instructions "0" or "PB." The instruction to pass to the power-supply test 0.56.0000.7747.0000 is sent to register Y according to these conditions, while in address 7766 of header is the instruction 0.56.0000. Y+1.0000. A 1 in position 2 of NVM-3 gives cycling of the example for constant RA (without influence on "0 absent" AA).

Test ω tests the correct operation of the tag ω circuits in arithmetic and logical operations. The test contains two groups of examples: with $\omega=1$ and $\omega=0$. Correct execution of an example is checked by instructions operating from the tag ω (instructions "36" and "76"). In the case of an error in one of the examples in the first group ($\omega=1$) halt occurs in register 0017, and in examples of the second group, in 0032. In both cases the example instruction is sent to R1 and the result to R2. The number of the instruction lights up in RA. After the examples of both groups have been executed the generation of tag ω by instruction "47" is tested individually. In the case of error halt occurs at address 0044 (for $\omega=1$) or 0051 (for $\omega=0$). In the individual mode the test role of positions 1 and 2 in NVM-3 is the same as in the test "0" absent." The same relates to the contents of NVM-2.

The test AA-1 tests the operation of AA in two ways: by a variable code which is obtained by suitable processing of the contents of NVM-1 and NVM-2, and by a constant code (five test examples are executed). In testing AA the variable reference code is formed in MA and the result in AA. In the case of error halt occurs at address 0025. The reference is sent to R1 and the result is formed in AA. In the case of errors with constant code halt occurs at address 0040 with output to R1 of the reference and to R2 of the result. In the individual mode the test of "1" in the 1st position of NVM-3 provides execution of the power-supply test before each test, "1" in the 2nd position provides execution of the first example with constant terms.

The control test checks the correctness of instructions "52" and "72," all control transfer instructions, and transfer after loops. The test consists of several examples, in which the correct fulfillment of all functions of these instructions is tested. In the individual mode "1" in the 1st position of NVM-3 provides execution of the power-supply test before the test. The code 0.56.0000.Y.0000, is set up in NVM-2, where Y is the address of the first instruction of the example with which the test begins.

The multiplication test checks correctness of the highest- and lowest-order bits of the product for different variants of the multiplication operation. It contains examples which are sequentially executed. Each example contains five words: a) the multiplication instruction of the example, b) the first factor, c) the second factor, d) the reference word of the highest- and e) of the lowest-order bits of the product. When the highest-order bits are incorrectly formed halt occurs at address 0010, and the lowest-order bits, at address 0012. In both cases the reference word is sent to R1 and the result to R2, and the number of the example lights up in RA. In the individual mode the test of "1" in position 1 of NVM-3 provides execution of the power-supply test before each test, "1" in position 2 of NVM-3 provides cycling of one example with power-supply test, "1" in position 3, execution of power-supply test before each test example.

The overall test checks execution of all variants of the arithmetic and logical operations. The test consists of several examples, each of which contains four words: a) the example instruction, b) the first number, c) the second number, and d) the result reference. When error occurs in the execution of an example halt occurs at address 0007 with output to R1 of the reference and R2 of the result of the example. The number of the example lights up in RA. In the individual mode "1" in position 1 of NVM-3 provides execution of the power-supply test before the entire test, "1" in position 2 provides cycling of a single example with power-supply test, and "1" in position 3, execution of the power-supply test before each example.

The test AU-1 checks operation of AU with a variable word which is generated by suitable processing of the contents of NVM-1 and NVM-2. The test contains two examples: a) execution of the operations of multiplication and division, b) multiplication and square root. If in either example machine zero is formed during multiplication, the examples are executed with the test constants as the initial numbers. On error in the first example halt occurs at address 0035, in the second, at address 0061, with zero sent to R1 and the result of the example to R2. In the individual mode test of "1" in position 1 of NVM-3 provides execution of the power-supply test before the first example, "1" in position 2 provides execution of the power-supply test before the second example, "1" in position 3, execution of the examples with constant operand words, "1" in position 4, cycling of the first example.

Test EA checks the operation of EA with a variable word obtained from NVM-1 and NVM-2 by the operations "07" and "54." With correct execution of the test the result in Ad should take the form EA, EA, ..., EA, where EA is a nine-bit code. In the case of error the nonets in Ad will differ, and the program stops in address 0031, with output to R1 of the result and to R2 of the reference. In the individual mode test of "1" in position 2 of NVM-3 provides execution of the power-supply test before the test EA, "1" in position 2, cycling with constant operand codes.

These tests in a somewhat modified form constitute the combined test.

The test of cyclic operations checks execution of instructions TsS, TsV, and SdTs with a variable word taken from NVM-1 and NVM-2. The test consists of two examples: a) check of the complex of operations TsS and TsV, b) check of SdTs. On error halt occurs at address 0017 or 0024 with output of the reference to R1 and the result to R2. Because of the properties of operations TsS and TsV, at least one button must be pushed in in NVM-1 in the exponent and mantissa positions. In the individual mode "1" in position 2 of NVM-3 provides execution of the power-supply test before the entire test, and 1 in position 2 of NVM-2, execution of the test with constant operands.

Test AU-2 checks AU for execution of all arithmetic and logical operations with a more complex regime due to the combination of operations. The test contains a large number of examples, each of which consists of four words: a) the instruction of the example, b) the first number, c) the second number, d) the result reference. The test fetches the instruction of the example and writes it in core addresses 1701-1777, where the third address of the instruction must be equal to 4601-4677, respectively. Control is then transferred to the start of execution of the examples in address 1701. After execution of all examples the correctness of the results is checked. In the case of error in any given example halt occurs at address 0026 with output to R1 of the result of the example and to R2 of the reference. The address of the subexample of the example (equal to the two last numbers in A3 of the subexample instruction) appears in RA. When the button "start CSC" is pushed in halt takes place at address 1700, with output to R1 of the example instruction and to R2 of the result. The address of the example appears in RA. Putting the machine in the mode "single-cycle" and pushing "start CSC," we exit to the instruction of the incorrectly executed example. In the individual mode the control test is performed from NVM-4. One in position 10 permits operation with all of core. The example instructions are stored in addresses 1701-4600, and the A3 of these instructions ranges between 4601 and 7500. With "1" in position 44 of NVM-4 exit to header on test termination is blocked, i.e., the power-supply test is not carried out before AU-2. The test provides for cycling of a single or of several examples, the so-called truncation. This mode is prescribed by "1" in position 45 of NVM-4 and the contents of positions 13 to 23 of NVM-4, the initial RA of truncation, and in 1 to 11 of NVM-4, the final RA of truncation. If they are equal, cycling takes place on one example.

The repetitive test of CM with constant code checks a single memory address repetitively. The test consists of two independent parts: the first tests registers 0400-7777, the second registers 0001-3400. The cube is checked by cartridges. The cartridges of CM are first erased by physical "zeros," and then the contents of the tested registers are checked for destruction of the initial information by "tickling." When individual bits "overoscillate" halt occurs at address 0026 in the first part of the test and at 3720 for the second part, with output to R1 of the word in the tested address and to R2 of the reference. The address of the defective register appears in RA. As a result of this the registers are repeated with physical ones by instructions 0.55. $A_i.A_i.A_i$ and 0.75. $A_i.A_i.A_i$. If individual bits are "hollowed out," halt occurs at address 0051 (for the first part) or 3742 (for the second part) with output of the contents to R1 and the reference to R2. The address of the defective register appears in RA. Each part of the test operates four times: twice by instruction "55" and twice by "75." The best operates only in individual mode and is controlled from NVM-4. The contents of positions 13 to 23 define the number of times the repeated section is run over; "1" in position 44 gives the allocation and repetition of CM by inverted words.

The test CCO MM tests the central control of MM, specifically, it checks the operations of access to MM for various conditional numbers in instruction "50." The test also checks the MD channel counter, correct zone selection on MD and MT, and the buffer loading and printing modes. Various forms of reversal are checked in operation with tape. In case of error halt occurs on the instructions stop or Aut. STOP by instruction "70." In individual mode the test is controlled from NVM-4. A 1 in position 45 provides buffer testing, in 44 test of tape MT-1, and in 43, of MD-1.

The repetitive test of CM by variable code is the same in structure as the ordinary test of CM by variable code with inclusion of "repeated" sections by instructions "55" and "75." The variable code is formed by cyclic addition of the constants of NVM-1, NVM-2, and NVM-3. Addresses 0061-7500 are tested. With overoscillation of bits halt occurs at address 0022 and with hollowing out, 0033. In both cases the reference is sent to R1 and the word in the tested register to R2. The tested address appears in RA. The test checks CM in the writing and regeneration modes, which is evident from the contents of register 0010. If $\langle 0010 \rangle$ = 1.75.0054. 0000.0000, then before repetition writing occurs in the tested register, if $\langle 0010 \rangle$ is equal to zero, writing is not carried out before repetition (test of code storage). The test operates only in the individual mode and is controlled from NVM-4. The complementary code of A2 in NVM-4 defines the number of times the repetition sections are run over on condition that A1 of NVM-4 is equal to zero.

The test of MM permits checking drums, tapes, output in variable code, formed by cyclic addition of the contents of NVM-1, NVM-2, and NVM-3. In cyclic mode the test MM operates like the ordinary CM test with variable code. Test MM is only possible in individual mode. On error in the write–read circuits of drum or tape Aut. STOP occurs by instruction "70" and after the button "start CSC," halt at address 0014. The code read from MM is sent to R1, the reference word in CM to R2. The contents of RA define the location in MM at which the error has occurred. Correctness of printing is checked visually, and of punching by fictitious input. In individual mode the test is controlled from NVM-4. A 1 in position 45 includes MM in the operation. Positions A1 of NVM-4 give the relative number of the instruction M_a (operating mode of MM), A2 the initial address of the MM zone, A3 the number of words in CM. Aside from this, "1" in the 43rd position of NVM-4 provides cyclicity of the test in access to MM and a "1" in 44 serves to modify the zone number of MM. The positions A2 give the step in zone number modification.

Test MD-1 checks the state of the write–read circuits of MD, magnetic heads, and magnetic coating by multiple writing in a single zone of MD. The test checks MD by parts, each of which contains 200_8 zones. If a defective section of the zone is found, halt occurs at address 0047 with output to R1 of the code read from MD, and to R2 of the reference. The test operates only in individual mode and is controlled from NVM-4. Positions A 1 give the number of MD and the presence or absence of BPC; "1" in position 44 prescribes writing 1 over 0, and "0" in the position, 0 over 1. Bit 45 must be zero. If bit 45 is 1, MD is tested by the code assembled in NVM-1, i.e., an arbitrary word. Positions A 2 of NVM-4 define the number of repetitions of writing in a given location of MD, where bits 13-18 define the number of repetitions in direct code and 19-24, in complementary code.

Test AA-2 checks the operation of AA with a constant word. It consists of two parts, located in different sections of CM (0001-0361 and 6300-6610, respectively). The test consists of several examples, which contain the following words: a) the RA instruction of the example (characteristic example) — first component — b) the TsS instruction of the example — second component — c) the instruction for transfer of the result of the example, d) the instruction for transfer of the initial number to the working register of the example. The positions of AA are tested for various combinations of 1 and 0. The example is executed in the following way. The reference is first set up in AA by cyclic addition 340_8 times by the instruction 0.07. $A_i.A_i.A_i$. The initial number for this instruction is equal to 0.00.0000.0000.0001. The TsS instruction of the example is then written in 340_8 successive CM registers. It has the form 7.07. $A_j A_j A_j$. Control is then transferred to the first of the TsS instructions of the example. The example forms $A_{eff} = A_j + RA_{example}$ in AA. If an error occurs in the functioning of AA, the result of the example is false and halt occurs on instruction "35" in register 0045 with output of reference to R1 and the result to R2. The number of the example will appear in RA. After the button "start CSC" is pushed halt occurs on instruction "17" at address 1000 with output of the TsS instruction of the incorrectly executed example to R1 and the contents of RA during this execution to R2. In the individual mode the test of "1" in position 1 of NVM-3 provides cycling of the first half of the test, "1" in position 2 gives cycling of the second half, "1" in position 3, operation of the test with NVM-3. In this mode positions 25-34 of NVM-3 enable the example to be written in an arbitrary number of registers, defined by the contents of this position, rather than in 340_8 registers.

The output test checks the operation of the printer and punch, operates only in individual mode, and is controlled from NVM-4. Positions A1 of NVM-4 give the output mode and have the following significance: "0" in A1 gives decimal printing for 120 lines, the first ten lines are decimal, the remaining positions of the mantissa and the 1st position of the exponent contain "1." In the 2nd position of the exponent appear 0, 1, 2, 3, 0, 1 ..., in the sign position is printed ---, +++, ---, +++, ... With "1" in the 25th position of NVM-4 octal lines are printed, "1" in the 26th, octal septets, "1" in the 27th, an octal "ladder" (the entire buffer is output). Each line has five "triplets" coinciding with the number of the buffer zone. With "1" in the 28th position 100 zeros are output in decimal print, in the 29th, decimal lines, in the 30th, decimal print of 192 words of the form: 64 lines of --- 37 777777777, 64 lines of ---3---------, 64 lines of ---3. With "1" in the 31st position a decimal "ladder" is printed. The entire buffer is output. Each line contains in its numerical part three identical triplets. Each of them is equal to the number of the buffer zone in decimal code. A "1" in position 32 of NVM-4 prescribes the "punching proper" mode; "1" in position 33 the decimal print of 100 words of the form: in the sign and exponent positions ---3, and in the numerical part minuses along diagonals; "1" in position 34 punches "moving zeros and ones," "1" in the 35th position, octal print of "moving zeros and ones," "1" in the 36th "test of ARep for punching."

The Aut. STOP test is intended for checking all Aut. STOP circuits of the machine for arithmetic operations and incorrect prescription of the operating mode in instruction "50."

After repeated performance of the complex test CT-3 in the preassigned limits to establish the effectiveness of the machine, the check tests appear. By their means the operation of the machine is tested as a whole, beginning from data input and ending with output of the results. They are taken from among ordinarily solved problems which for one or another reason suffer perturbations during their solution, regardless of good performance in the tests. Most often this is a problem which heavily loads CM (repeated access to certain addresses) or requiring complicated operating modes of AU (problems in the solution of differential equations by Euler step methods, Runge – Kutta, etc., of algebraic equations by iterative methods, etc.). Aside from this, problems are encountered which change the frequency of machine operation due to repeated accesses to drums and tapes. If the check problem is perturbed, the circuits in which the fault occurred are carefully examined. When the check problem is solved correctly, the machine is authorized for computation. The check problems are changed from time to time.

§2. Fault Analysis in the Execution of Individual Instructions

Let us consider certain aspects of fault localization in the execution of individual instructions in various modes ("individual," "single-cycle," etc.) with cycling of a single example in the test or cycling of an example from the control console registers.

The individual mode permits the character and possible cause of failure to be detected, and more rarely the cause itself; further localization of the site of failure is carried out by means of technical equipment. Let us elucidate this by examples.

Example. On execution of the instruction 000.7777.0000.0000 in the individual mode incorrect formation of A1 occurs on the pulse CSC P3; in place of all ones in the address adder the code 7767 is formed.

The possible cause is one of the elements in the circuits of the fourth position in AA. To localize the fault we interchange the adder blocks between the 4th and 5th positions. On repeated execution of the instruction the error has shifted to the 5th position.

Example. On execution of the instruction 004.7771.7772.0000 in individual mode, where $\langle 7771 \rangle = 777.7777.7777.7777$ and $\langle 7772 \rangle = 101.4000.0000.0000$, on pulse P10 there appears in MA the word 3773.7777.7777 in place of the word 3777.7777.7777.

After replacement of the paraphase flipflop in position 27 of register R2 the fault was eliminated. Indeed, MA should contain

$$+\frac{\begin{array}{c}7777.\ 7777.\ 7777\\ 3777.\ 7777.\ 7777\end{array}}{3777.\ 7777.\ 7777}$$ (complementary code to 4000.0000.0000).

If the paraphase flipflop in the 27th position of R2 had no output, then

$$+\frac{\begin{array}{c}7777.\ 7777.\ 7777\\ 3773.\ 7777.\ 7777\end{array}}{3773.\ 7777.\ 7777}$$ (complementary code to 4000.0000.0000, except for the 27th bit).

However, it is not always possible to find the cause from the nature of the error. Let us give an example.

Example. On execution of the instruction 0.01.0100.0275.1000, where $\langle 0100 \rangle = 101.0000.0000.4000$, and $\langle 0275 \rangle$ is equal to zero, in the individual mode (with bias −43 V, it was found that the code was not transmitted from the 12th position of R1 in MA at the 12th pulse of CSC.

By use of the oscilloscope it was possible to find the true cause of the error: "underheating" of the tube in flipflop F "0" of R2 and its output was very small. Indeed, since the second number is "0," on P11 CSC F ω should pass to "1," and its output conditions processing of the signal +No1 (P12, "1" F "ω" +−|−|, cf. Table 3, Section 5.2). The flipflop Fω does not always arrive in "1," since F0 in R2 gave a weak intermittent signal when set to one.

In the single-cycle mode the instructions are tested more infrequently than in individual and automatic modes, and often the single-cycle mode is used for transfers or change of instruction, transfer of control, etc. We present three examples of application of the single-cycle mode to test correctness of execution of instructions.

Example. On execution of the instruction 0.05.0100.0375.1000 in the overall test CT an intermittent failure occurs at −44 V. The contents of the register in A2 are found to be equal to ⟨0365⟩ of the register.

We test instruction fetch to the instruction register, for which: a) we block CRA, b) we put the working mode switch in the position "single-cycle," c) we change the bias to −45 V, d) we erase the instruction in IR from the reset button of the executive register and we again pull out the button "start CSC." Working in this mode (erasure − start) we note that sometimes in place of the given instruction the instruction 0.05.0100.0365.1000 is fetched. The possible cause of the error is in the 16th position of the CM-IR circuit.

Example. In execution of the instruction 0.15.0101.0115.1000 an incorrect result is obtained in register 1000, equal to ⟨0115⟩ shifted one place to the left.

Operating in the sequence indicated for the preceding example, we note that sometimes the instruction "14" appears in the operation register instead of the required one.

Example. In execution of the instruction 0.15.7771.7775.0000 where ⟨7771⟩ = 7.77.7777.7777.7777 (with toggles Bl NVM, Bl IR, and Bl CRA) we test AM and the code bus amplifiers A3 in cyclic mode. At the first bitwise addition the result should be all binary ones, on the second, zero, on the third again all ones, etc. The faulty positions are easily observed on the neon lamps of CC.

The automatic mode permits the location of an intermittent failure to be determined when the test does not give results in the individual or single-cycle modes.

Example. On execution of the instruction 0.07.7771.7772.0000 in the automatic mode, where ⟨7771⟩ = 777.7777.7777.7777 and the contents of 7772 are a word obtained by successive depression and raising of each of the 45 positions of NVM-2, we obtain in RR a word consisting of zeros and ones when the k-th position is pushed in, which easily permits fault detection in the corresponding position of AM or CM.

Example. As in the execution of instruction 0.27.7771.7772.0000 it is fairly easy to detect failures in the paraphase flipflops of register R2.

With successive depression and raising of the positions in NVM-2 we obtain in RR one "floating" hole.

Example. On execution of the same instruction with ⟨7771⟩ = 777.7777.7777.7777 and ⟨7772⟩ = 777.7777.7777.7776 (1st position of NVM-2 raised), we obtain the word 0.00.0000.0000.0001 in RR. If we raise the 2nd position of NVM we obtain "1" in the second position of RR, etc., which permits faults in the paraphase flipflops of R2, AM, and CM to be detected in the corresponding positions.

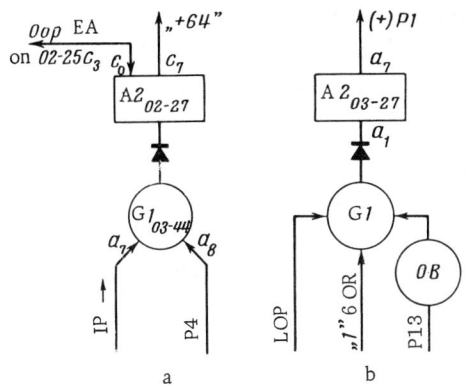

Fig. 1. Circuit changes in instruction "20."

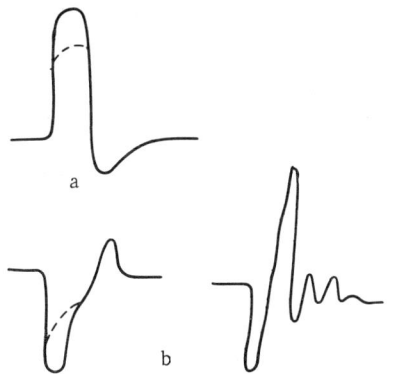

Fig. 2. Change of signals at the check points of elements F_z: a) at point a_0; b) at point a_2.

The automatic-single-cycle mode (start-stop mode) sets up a dynamic load on the machine elements taking part in the operation (repeated execution of the example, pause, again repeated execution, pause, etc.). We give two examples of the use of this mode.

Example. In the repetitive test of CM halts occur with RA = 2216.

After restoration of the contents of register 2116 in CM by instruction 3.20.0001.0000.0000,* where $\langle 7771 \rangle$ = 7.77.7777.7777.7777, we execute the instruction 7.55.0000.0000.0000† "in automatic-single-cycle" mode. On execution of this instruction in the given mode several bits have again "dropped out" in MA (all bits can drop out). The location of the failure may be found by interchanging the corresponding elements F_z or examination of the check-point signals by oscilloscope. Figure 2 shows the changes in waveform and the magnitudes of the pulses in the elements F_z when tested in the start-stop mode.

Example. In the automatic-single cycle mode the effectiveness of blocks CRA, RA, and AA are usually tested using the instruction 2.52.0000.0001.0000 (or 1.12.0000.0000.0001.). At halt the positions of CRA, RA, and AA must coincide (the keyboard APU is reset, and flipflop Bl IR set).

On disagreement in any position it is possible to define approximately the location of the failure of the corresponding position of AA, CRA, or RA.

Examination of examples leads to the following conclusions:

1. There exists a relatively small set of examples (test examples, or set up on NVM) which permit a large number of machine failures to be determined directly from the control console using various operating modes of the machine.

2. Logical procedures for detecting failures have a standard character and are easily remembered.

---

*For convenience of detecting faults from the control console the operation PR2 ("1" 5 OR, 00, P9; Table 3, Fig. 1) is added in instruction "20." This permits reading a word from the same address in which writing has occurred without resort to other instructions for this.

†Since VAPU and PRR occur on pulses P7 and P10, the word-fetch register at the control must be erased. For instructions "40" and "60" the output of register R1 is blocked (00, "0" 6 OR, P17).

3. The machine failures considered in §§1 and 2 may be divided into typical and atypical (in greater detail see below). Typical failures are those which are met more or less frequently in the same circuits (blocks). Atypical are failures which have a relatively rare, random character.

4. Logical analysis most often permits the circuit in which a failure is expressed to be found. Further localization of the failure must be carried out by technical means.

CHAPTER II

Failure Detection by Technical Means (technicological analysis)

If a failure cannot be eliminated by logical analysis, then technical means are used: oscilloscope, measuring instruments, various auxiliary circuits, etc. This method offers the possibility of localizing intermittent failures, of analyzing faults in all circuits and blocks of the machine, of checking the form and magnitude of the signals at test points, at the outputs of the elements, etc. If logical analysis permits the character and location of the failure to be estimated only approximately and to eliminate it by successive search, the technical means also help to grasp the cause of the failure. Logical analysis requires much less time to eliminate simple failures and with accumulated experience predominates over, but does not replace technico-logical analysis.

§1. Choice of Synchronization Method for the Localization of Intermittent Failures

The basic method of search and elimination of failures using technico-logical analysis is the cycling of instructions during whose execution the failures occur. In more complicated cases cycling of groups of instructions is used to localize failures. Certain auxiliary circuits, described below, are used for the choice of synchronization. The majority of synchronization points are brought out to the autonomous control console (ACC) – the CSC, A "1" K, "3072," series, etc., pulses. The concrete choice of synchronization depends on the device whose circuits are being analyzed (CM, AU, CU, MM, peripherals). General procedures for choice of synchronization can be given according to this.

To examine the CM writing circuits it is convenient to take the synchronization from P16 of CSC, in order to "tie in" to the instant of writing in CM. In analysis of the CM reading circuits synchronization is taken from P5 or P19 of CSC (depending on the address – A1 or A2 – for which the writing circuits are analyzed). It is convenient to examine the control circuits of CM using synchronization from one of the reading pulses (R) and the address decoders using the pulses A "0" of CM.

The choice of synchronization for the analysis of AU operation is more complicated. As is well known, AU executes two forms of operations: 1) those which enter into the standard operating cycle of the machine (logical, cyclic, etc.), and 2) not entering into the standard cycle (multiplication, division, etc.). If intermittent failures occur in AU blocks (R1, R2, MA and EA, KShCh, RR, AUA) on execution of the first type of operation, synchronization is taken from the CSC pulse determining the instant that the reception signal is generated (PR1, PR2, etc.) or the addition signal (+No1, (+)No1, +No2, etc.), depending on the character or the failure. More often failures arise in the execution of the second type of operation, since the AU blocks work in a more difficult regime. In this case, aside from the earlier considerations, in the analysis

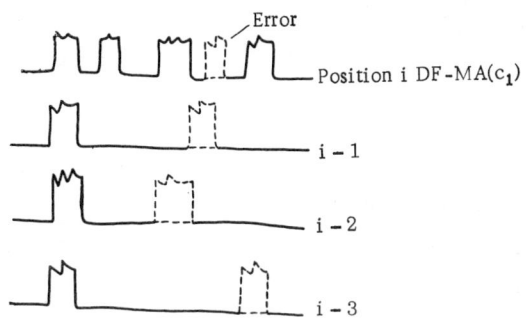

Fig. 3. Shift of failure time in the analysis of test points c_1 in flipflops of MA.

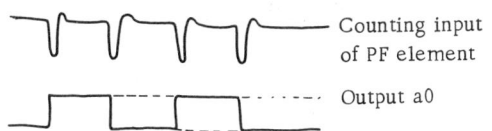

Fig. 4. Errors in counter "512" flipflop.

of blocks R1, R2, MA and EA, and the local control blocks of the operations, synchronization is taken from the pulse of CSC which determines the instant that the corresponding LCO circuit is triggered.

In the analysis of intermittent failures in the blocks for generating the elementary operations CCO CU synchronization is taken from the CSC pulse preceding the failure. The corresponding pulses are brought out on the panels of CU or MMCU to synchronize the MM circuits. The choice of the concrete synchronization pulse depends on the tested mode (starting period of MM, search for zone number, writing or reading words, termination of MM operation).

In checking input devices (the test mode is defined by the toggles RKK, SP, OM, and AM) P3 of CSC, SPIK, or OM and AM are used for synchronization. The output control circuit is tested with synchronization by pulses "3072," NPCh, or KPch, and in search for more complicated intermittent failures, from preceding effective circuits.

Let us give some examples of the choice of synchronization.

Example. In execution of the instructions 0.04.7771.7772.0000, where $\langle 7771 \rangle$ = 7.77.7777.7777.7777, $\langle 7772 \rangle$ =1.01.6000.0000.0000, failures occur in the mantissa adder. Find the location of the failure.

Synchronization is taken from the 10th pulse of CSC, since the circuit LCO Div is triggered by P10. In the analysis of the signals at the test points C_1 of the MA flipflops a shift of the failure time is apparent (Fig. 3.1). In the analysis of the failure location in the given position (i − 2) signals are examined at the test points of AM (a_5, a_8, c_4, etc.), CM (c_8, a_2, a_6, etc.), DF (c_1, c_6, etc.), and at the inputs and outputs of amplifiers A2.

Example. At −43 V counter "512" displays breakdown. Only zeros are printed.

The possible cause is transient failures in the printer counter. To analyze the operation of the printer counter "512" synchronization is taken from the first SIB (the pulse train "3072" of the printer is fed from ACC to the synchronization input of the oscilloscope over a diode), or from the signal A "1" K from ACC, or from start of print, if the failures are taking place in the flipflop "1" K and the pulse train "3072." The flipflop output signals are easily inspected at the points a_0 and c_0 (Fig. 4).

§2. Utilization of Auxiliary Circuits for the Localization of Transient Failures

A circuit is given in Fig. 5 which permits transient failures to be located. The circuit has been developed in connection with the problem that in certain modes, connected with changes in the working frequency (for example, in writing IS-2 on magnetic drum and reading with check), the signal $\omega = 1$ is not generated on the instruction 15 A1.A2.A3, with the result that the machine begins to "cycle." It was not possible to detect the site of the failure by ordinary means, since with the usual execution of this instruction the signal $\omega = 1$ was generated in the presence of arbitrary changes in −40 V.

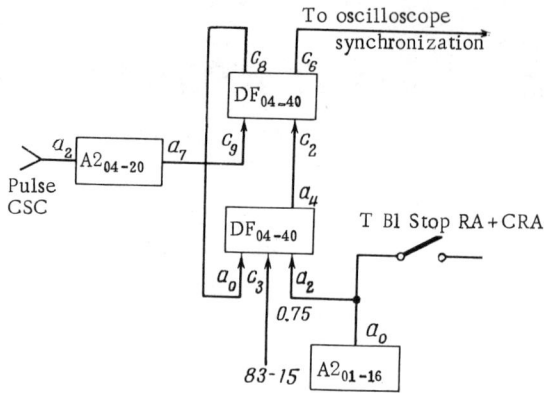

Fig. 5. Circuit for transient failure localization.

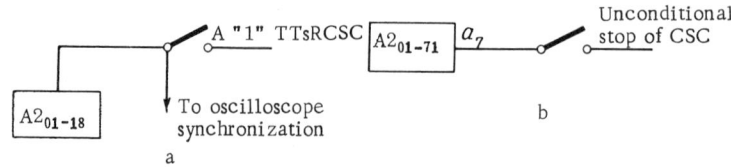

Fig. 6. Inserting toggle in circuit (a) A "1" TTsR CSC and (b) unconditional stop.

The circuit operates in the following way: oscilloscope synchronization is taken at the output of $DF_{04-40}c_6$, whose setting to "1" is realized from the output of the second half of the flipflop element. CM contains a set of instructions describing a certain number of arbitrary words on magnetic drum and reading them with check by the instruction 15 A1.A2.A3. The register number in which the comparison instruction is located is set up on the keyboard of CRA and the toggle "stop on CRA+RA" is set on CC. The machine stops by toggle "B1 stop RA and CRA" on ACC are removed. In this way, each time that a comparison instruction is executed a machine stop signal is generated which sets flipflop DF_{04-40} to "1" (Fig. 5). The oscilloscope timebase triggering is bound by the circuit to any pulse CSC, so that by running over these pulses it is possible to localize the failure. Using this circuit the failure was eliminated in several minutes (one of the amplifiers A2 in the circuit "0 absent" MA had intermittent failures). The circuit can also be used to locate other failures.

The circuit shown in Fig. 6a was developed to extend the possibilities of synchronization. As is well known, on coincidence of the address of writing in some register of CM with the keyboard address "Stop on Wr" with toggle "stop on Wr" set in the mode of reading from drum or tape (instruction Md "1" R) stop CSC on P4 occurs (the signal A "1" TTsR CSC). If a toggle is inserted at the output of $A2_{01-18}$, interrupting the signal A "1" TTsR CSC, in the presence of these conditions conditional stop of CSC does not occur, and at the instant of coincidence the oscilloscope timebase is triggered. This circuit was made because in the writing of certain blocks of numbers on magnetic tape it was not known for what reason information was lost in some parts. Since these locations varied very greatly it was assumed that the information is lost as a result of some kind of noise or parasitic pulses in the electronic apparatus. The examination of the circuit using the toggle in the network A "1" TTsR CSC permitted the oscilloscope time base to be triggered at the failure points by progressive approach to these parts using the keyboard Stop on address.

To localize sources of parasites in checking the repetitive circuits a toggle is inserted in the circuit of unconditional stop of CSC which removes the Aut. STOP mode (Fig. 6b). This appreciably helps the search for local parasitic failures due to automatic stops of the machine and reduces the search time.

The failure points in the output punch (determination of punch grid skew) are localized by program. The test program "input-output" takes 500_8 CM registers and punches out the contents of registers 500_8 to 1463_8 (i.e., 763_8 registers). In the CM registers beginning from 1501 is written the following program:

```
1501  6. 35 0500. 1504. 0000
1502  1. 12 0763. 1501. 0001
1503  0. 17 0000. 0000. 0000
```

This program makes possible:

1) comparison of the contents of CM registers (reference) from 500 to 1463 with the information punched out, which is written in CM beginning from register 1504 (immediately after the program) and ending at register 1567;

2) determination of the place of skew (the card, and number on the card) according to the number in the address register (RA). For example, in testing punching the machine stopped with the contents of RA at that moment equal to 50_8-40_{10}. Since there are 12 rows on a card, therefore (40 : 12 = 3, with remainder 4), three cards had been read, stop occurred on the fourth, reading the fourth row from the top.

The use of special diagnostic programs and auxiliary devices (testers, generators, etc.) is not described in this article.

Let us consider the forms of preventive maintenance.

§3. Preventive Maintenance

The preventive maintenance tests of the computer include the following types of work.

A. Marginal testing.

1. Testing CM without repetition on variable code: a) with bias of AR-2 over the limits -55 to $+15$ V, b) with flipflop misalignment ± 6 V, c) with change of $+300$ V from $+280$ to $+320$ V (test for "tickling" of F_x), d) with variation of the voltage $+200$ V by ± 15 V, e) with variation of the base voltage -40 V from -34 to $+45$ V.

2. Testing CM with repetition and reducing the anode voltage to $+185$ V, change of -22 to -23 V and -40 to -41 V.

3. Testing AU and CU by the combined test and the test AU-2 with changes of -40 V between -35 and -45 V.

Testing magnetic drums by MM with variable code with change of bias in the basic accumulator over the range -25 V to -50 V and CCO MM from -35 to -45 V.

5. Testing the magnetic tape blocks with variation of -40 V between -35 and -45 V.

6. Testing the printer by the print test with "rolling" in the regeneration amplifier voltage between -25 and -50 V and the basic voltage -40 V between -35 and -45 V.

7. Test of output punches: a) with the "input-output" test by the 32nd and 36th positions of NVM-4; b) checking the output deck of punched cards by mask to detect possible misalignment; c) testing output on CR in the top and bottom blocks; d) testing output of the output punch for doubling and checking doubling by mask and CR.

8. Checking slow print from the output of rows to the output punch by the MM test.

The operation of the machine is considered to be stable if no transient failures occurred with variation of the voltages in the indicated limits.

B. Testing the machine by the check problems (Chapter I, §1).

C. Checking the power elements of the machine.

To prevent possible failures and breakdowns the power elements of the various networks are tested according to a schedule. Several variants of these checks are possible.

1. Test by operations. All power elements (amplifiers, flipflops, additions and carry blocks, etc.) taking part in a given operation are checked.

2. Test of the selected networks in the working mode. The output signals of CSC, OR, OC, R1 and R2, the accumulator, address part, etc. (the most heavily loaded circuits).

3. Test of power elements by circuit boards. First the pulses of the working train, CSC, OC, and OR, the addition and carry block test points, the outputs of register and accumulator flipflops are tested, and then the power elements by boards. One of the boards is examined with correct biasses. The preventive maintenance is carried out by special procedures which indicate the set of instructions and board elements taking part in their execution. The cycle is closed over 40 days and the analysis of the working circuits of the machine begins again.

4. The advantages and defects of each method. A drawback of testing by operations is that for each new instruction circuits are tested which have already been examined in the preceding operations. Such duplication increases the time of preventive maintenance (the cycle is closed after 60 days). Other drawbacks of this procedure are also well known: in the addition operation, for example, it is necessary to take into account the states of various elements in dependence on conditions (equal exponents, first number greater, second number greater, one of the numbers zero or negative, etc.) which, as with duplication, substantially lengthens the preventive maintenance time. Therefore this variant is hardly acceptable, although it has its advantages: clarity, great cognitive value.

A drawback of testing selected circuits is that not all elements are examined, and therefore there always remain locations which could be the cause of certain random parasitic signals.

The third variant combines the best features of the first two methods. Its drawback is its magnitude and the difficulty of prior preparation.

CHAPTER III

Typical Failures of the Basic Devices of M-20

§1. Typical Failures of CM and Their Elimination

The typical failures of CM include failures in the F_z elements, in the word write and read circuits. Usually they appear during the period of preventive maintenance while determining the working margins by the tests. The operating margins of the machine are found by the CM test without repetition for the basic voltage -40 V. With variable codes in NVM the test must be completed without failures in the range -34 to -45 V. A preliminary test of CM by this test immediately reveals its basic failures (write or read circuits, CM control, etc.), whole localization is realized in stages.

1. At the basic voltage -40 V the reading amplifier AR2 voltage is varied from ACC in the cutoff direction to -55 V, which permits the weakest of them to be eliminated. With opening to $+15$ V the reading amplifiers giving false signals are eliminated. After this test normal bias -20 V is restored to AR2 and the number register flipflops are tested.

2. Failures of number register flipflops are easily found by the CM test without repetition by misalignment between ones and zeros using the toggle on ACC CM. At misalignment the address decoder flipflops are also tested. To find a defective position PF of the address decoders the address code set up in NVM-1 and NVM-2 is

$$\langle \text{NVM-1} \rangle = 0.\ 00.\ 0032.\ 0032.\ 0032,$$
$$\langle \text{NVM-2} \rangle = 0.\ 00.\ 0001.\ 0001.\ 0001,$$
$$\langle \text{NVM-3} \rangle \quad \text{equal to zero.}$$

3. The writing amplifiers F_x are checked by two tests: CM without repetition with variation of the voltage +300 V between +280 and +320 V and CM with repetition with reduction of the voltage +200 V to +185 V. An error on "overoscillation" of the k-th position is found by successively executing the instructions

$$5.\ 55.\ 0000.\ 0000.\ 0000,$$
$$7.\ 75.\ 0000.\ 0000.\ 0000$$

and in hollowing out by the instructions

$$3.\ 20.\ 0001.\ 0000.\ 0000,$$
$$7.\ 55.\ 0000.\ 0000.\ 0000.$$

4. Faulty diodes G1 and G2 in the read–write circuits are detected in the operation of test CM without repetition and pumping of the basic voltage −40 V between −34 and −45 V. The site of the failure is found by successive permutation of one of them (G1 or G2) from position to position.

5. The test CM without repetition permits failures to be located in the F_z elements also. If on variation of −40 V in these limits different positions drop out with constant values of the codes in the lowest or highest positions of RA (for example, RA-0555, 0573, 0521, etc.), the corresponding element F_z is changed (in this example $\Phi_{zy} = 05$). It is convenient to observe elements F_z at the points a_0 (tube 6P9) or a_2 (tube GU-50) using an oscilloscope. The instructions 3.20.0001.0000.0000 and 7.55.0000.0000.0000. are executed in succession at CC, where the code 7.77.7777.7777.7777 is picked up from NVM-1. On the first instruction the register defined by RA is filled with all ones, on the second repetition occurs. The repetition instruction is blocked in the start-stop mode.

In daily maintenance the test CM with repetition is run through two or three times. The number of repetitions is taken as 7-15. The most difficult operating conditions are set up, −41 V, +185, −23 V, the other potentials being nominal. Daily maintenance of CM with the repetition test permits the period of autonomous testing of CM to be lengthened to once a month (instead of four times or more).

§2. Typical Failures in Backup Storage (Magnetic Drums and Tapes) and Their Control Circuits

This type includes failures: (1) in the word read–write circuits; (2) caused by the magnetic heads; or (3) in the channel counter, in the reproduction amplifiers SP, zone number, P12B, and in the control circuit.

Let us briefly consider each type.

1. In the word write – read circuits (elements ARep, AWr "1," and AWr "0," A2, DF, matching the flipflops of register R1) a major portion of the failures occur in amplifiers ARep3 and write amplifiers AWr "1" and AWr "0." The amplifiers A2 and the dynamic flipflops DF at the input to the write amplifiers operate reliably and they are replaced not more than once or twice a year.

The variation of nominal -40 V at the reproduction amplifiers ARep3 of the basic store over the range -25 to -50 V permits defects to be eliminated in the corresponding positions. The location (position) is found by the test MM. A simple replacement or interchange of ARep3 does not always lead to the required result. Indeed, a weak write pulse current leaves a weak trace on the drum surface, and therefore even with a good read amplifier biased to -25 V the word may be poorly read. In this case the elements AWr-2 are changed to "1" in the corresponding positions.

Example. In testing drums with test MM and biasing ARep to -25 V "1" was not read in the 7th position (in various channels and drums).

The possible cause is the element ARep or AWr2 "1" in the 7th position. By interchange we verify that the element AWr2 "1" in the 7th position is faulty.

At high bias (to -50 V) the reproduction amplifiers can detect noise signals written on the drum by the amplifiers for writing "0" AWr2. Defects in these amplifiers are found analogously to the above.

The preventive maintenance of these blocks is carried out about once a month. The amplitude and duration of the detected signals at the test points c_1, a_0, a_9, etc., in ARep3, the output signals of the register R1 flipflops, the receiving amplifiers and flipflops in MM, the signals at the test points a_1, a_9 in AWr2, etc., are checked.

The defects in the magnetic tape write – read circuits are detected by the test MM, in an autonomous check, or in tape operation in the marking mode. The simplest defects in the different positions are eliminated as for magnetic drum.

The signals at the test points of Awr, ARep1, and ARep2 must satisfy the required values in all positions. The signals A "1" F "K" or "VTsK," obtained from ACC, are used for synchronization, or "self" triggering is used. The basic nominal voltages for the tapes are situated in the limits -10 to -12 V on SP, -15 to -17 V for the codes.

2. Defects in the magnetic heads are also typical failures. Their basic forms are: dirtying of the working surfaces of the heads, enlargement of the working air gap, defects in the heads. They are located by an autonomous check of the magnetic drums (at the test points of ARep3 in all channels) or by checking the drums by test MM with cycling in all channels in which they appear.

3. Prophylactic inspection of signals in the mass storage control circuits is based on the following conditions:

a) the control circuit must operate stably in the range -35 to -45 V for the basic nominal potential -40 V;

b) all signals must satisfy the prescribed values for amplitude and duration; they are checked once in 40 days by boards.

Let us present an example of a typical failure in a mass storage control circuit.

Example. The circuit A "1" K (setting to "1" of the word flipflop) does not operate.

This circuit is checked by means of the input device. For this the toggles RKK and SP are set, toggle AM is reset (all "1"s are written in CM). Synchronization is taken from the third pulse of CSC. The character of the failure is easily seen (Fig. 7a).

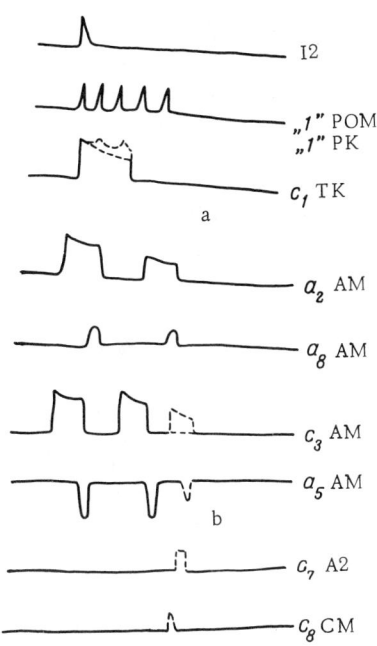

Fig. 7. Parasitic pulses (a) in flipflop A "1" K and (b) in element CM and AM.

Other typical failures occur in the circuit for generating "+1" in MD, where most often the monostable MS_{03-15} is disturbed. Amplifiers A2 are often disturbed in the channel counter. This failure is investigated by the test MM with variable code and repetitive writing or reading. The number of codes is such that automatic counting of the channels takes place. Synchronization is taken from the SPD pulses.

Defects in the reproduction amplifiers SP, zone number, and P12B are checked by variation of the nominal potential in them between -30 and -50 V with examination of test points a_9 in the elements ARep3. To count the drum zone numbers (synch-pulse zone No.) it is convenient to use the following procedure. The following sequence of instructions is set up in NVM:

$$\langle NVM-1 \rangle = 2.\ 50.\quad CN\quad 0000.\ 0001,$$
$$\langle NVM-2 \rangle = 0.\ 70.\ 0001.\ 7771.\ 0000,$$
$$\langle NVM-3 \rangle = 1.\ 12.\ 7777.\ 7771.\ 0001,$$
$$\langle NVM-4 \rangle = 0.\ 16.\ 0000.\ 7771.\ 0000$$

and the toggle BlNVM is reset. These instructions are executed in the automatic mode.

§3. Typical Failures in the Arithmetic Unit

The operation of AU is checked by means of test programs and check problems. With respect to the potential -40 V the operating range of AU is -35 to -45 V. Improvements and planned preventive maintenance make it one of the most reliable units of the machine. For example, the word bus amplifiers A3 in the arithmetic unit formerly failed. They substantially limited the working range of the machine with respect to the potential -40 V. Elements A3 were replaced almost daily. After change-over of the high-turns windings in the amplifiers A2 in CM with regeneration in the KShCh, and the low-turns windings by regeneration, the amplifiers A3 began to operate stably. Only 5 to 10 elements are ordinarily changed yearly.

Typical AU failures occur in: 1) the addition and carry blocks; 2) the flipflops Ad, R1, and R2; 3) the synchronization blocks. The remaining AU failures are not typical since they arise fairly rarely. The prophylactic inspection of each AU block (except those in which typical failures appear) is carried out by boards about once in two months.

1. The addition and carry blocks of AU are inspected about once every two weeks. One week, with instruction 0.04.7771.7772.0000 (and toggles T Bl IR, Bl CRA, Bl NVM reset), where $\langle 7771 \rangle = 7.77.7777.7777.7777$, $\langle 7772 \rangle = 1.01.4000.0000.0000$ or $1.01.6000.0000.0000$, and with bias -42 or -43 V the signals are checked at the test points a_5, a_8, and c_4 for the elements AM and a_6, c_8 for the elements CM, which permits their failure during a working shift to be almost completely prevented and to forestall possible transient failures.

With accumulator failures their location is determined in the following way. The bias voltage -40 V is set so that the character of the parasite is clearly seen. Usually the locations of parasites in the accumulator are detected by the division instruction 0.04.7771.7772.0000, where $\langle 7771 \rangle = 7.77.7777.7777.7777$, and $\langle 7772 \rangle = 1.01.4000.0000.0000$, or $1.01.6000.0000.0000$, or $1.01.7000.0000.0000$, or $1.01.7400.0000.0000$, etc. Then the result of division is all 1's or in MA 1's appear every other position or every fourth, etc.

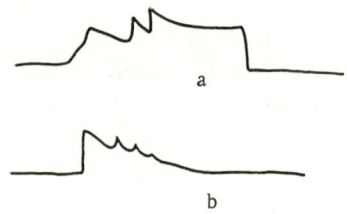

Fig. 8. Change of signals at the test points c_1 of the flip-flops Ad. a) Weak signal from U "1"; b) weak feedback.

The location of the failure is determined by oscilloscope at points c_1 of DF MA, starting from the 1st or 36th position. The position in which the parasitic signal begins earlier than elsewhere is inspected (cf. Fig. 3). Figure 7b shows an example of the location of the site of a parasitic signal and its cause.

The oscilloscope is synchronized so that the character of the parasite and its variation from position to position is clearly visible. For parasitic failures in the accumulator and the execution of the division instruction synchronization is usually taken from pulses P5, P10, P11, or P12 of CSC.

Let us present one further example. In execution of the instruction 0.01.7771.7772.0000, where $\langle 7771 \rangle = 1.37.7777.7777.7777$, $\langle 7772 \rangle = 1.01.4000.0000.0000$, parasites appear in the exponent accumulator. The character of the parasite is expressed very shaprly: it seems that CSC, the elements of the circuits LCO Sd, LC, etc., are being disturbed. The failure may be analyzed at the inputs to "spider" — a gate to which signals are applied from all positions of EA flipflops. A parasitic signal in any position, arising earlier than the others, is clearly visible. To locate the site of the failure in the given position the primary cause is investigated, which is disturbed earliest: AM, A2, DF or CM, etc.

2. The flipflops of register R1, R2, and the adder are checked in preventive maintenance about once in two weeks at the check points C_1. The signals at the flipflop check points and the prediction of possible parasitic signals in them are shown in Fig. 8. Defects in the paraphase flipflop circuits of register R2 are easily detected with division instructions (from P10), cyclic subtraction, by the instruction PTs, etc.

As has already been mentioned, the site of a parasitic signal in the adder is found by examining the signals at the flipflop grids. Similarly, in parasitic signals in the register R1 during right or left shifts the site of the parasite can be determined at the grids of the flipflops of register R1. The cause of failure in dynamic flipfops is a weak output due to weak tubes (loss of emission), punchthrough of the input diodes, parasitic signals due to "leaky" diodes, tubes, flashover in transformers, etc.

3. Defects in the synchronization blocks are prevented by planned preventive maintenance which is carried out about once a month for these blocks. The output signal amplitudes at all amplifiers A5 operating in series are checked.

§4. Control Unit and Its Typical Failures

The control device interconnects all units of the machine into a single complex, defining the sequence and nature of their operation. It is the most complicated, and the defects that arise in it are particularly difficult to trace, since their expression is very specific. Greater care in the preliminary analysis is needed than in the investigation of failures in the other units. The common feature of failures in CU is that the result obtained differs strongly from the reference. The basic typical failures in CU arise in: the block for formation of addresses, in the synchronization block, in CSC, and the circuits for pulse-train distribution, and in the blocks for generating the control signals for the operation of the other units.

The first two types of failure are met fairly frequently. Let us consider the failures in the following order: 1) the positional circuits of registers IR, OR, CRA, RA; 2) in the block AA (positional failures); 3) in the operation commutator OC; 4) in the synchronization unit; 5) in CSC; 6) in the pulse-train distribution circuits.

In localizing failures in CU by logical means the analysis is usually insufficient, and therefore they are used in conjunction with technical means. The best method for eliminating CU failures is to prevent them by planned preventive maintenance. During preventive maintenance the output signal and check point signal amplitudes are checked. This check is carried out about once a month.

The operation of the IR and OC register positions is checked in the following way: all 1's are set up in NVM-1, the unconditional stop toggle and the CRA blocking toggle are reset, the instruction 0.56.0000.7771.0000 is set up in IR, and the button "start CSC" is pushed to automatic. In this mode the input and output circuits of all positions in IR and OR are checked.

It is convenient to check the operation of the input and output circuits of the positions of CRA and RA in the execution of the instruction 0.56.0000.7777.0000 or 0.52.0000.7777.0000 with the toggles for blocking of CRA and IR reset.

The positional networks of AA (elements AM, CM, DF, and A2) are checked in the execution of one of the arithmetic or logical instructions included in the standard cycle. For convenient oscilloscopic examination and determination of the site of the parasite the effective addresses of the instruction must be equal to each other. In the instruction cycle it is first necessary to transfer the number to RA 0.52.0000.A2.0000. Then the tested instruction, for example, $7.07.B_i.B_i.B_i$, is executed. By varying the codes A2 and B_i it is possible to check AA in various operating modes. Synchronization for these checks is taken from the CSC pulses corresponding to the time of reception of the address addition signal.

To test RA, AA, and CRA in the complex from the control console, operation 2.52.0000.0001.0000 or 1.12.0000.0000.0001 is used in "start-stop" mode with blocked IR. In correct operation after the stop the codes in RA, CRA, and AA must be the same. Their disagreement indicates the presence of a transient failure. However, it is particularly difficult to determine the cause of failure using the oscillograph due to synchronization problems (in such a mode synchronization is taken from the position preceding the position being inspected). Usually the check by these instructions serves only to establish the fact of transient failure. The cause of failure is found by one of the earlier mentioned methods.

The OC networks are checked less often, about once in a month and a half, in standard manner: the codes of various operations are set up in IR, in the automatic mode. Particular attention is given in the examination to signals given out by the synchronization block (pulses 0.00.0.375.0.75 and 1.125), the pulse-train distribution circuits, and CSC, since the failures caused by them are not subject to analysis. Improvement of the synchronization and CSC has substantially improved the reliability of their operation with respect to the basic nominal potential − 40 V, so that at the present time the elements A2 and DF used in these circuits operate stably. The amplifiers A5 are inspected at reduced potential +300 V (to +280 V). Particular attention is given to the duration and amplitude of the A5 output signals. In the analysis of CSC failures a multivibrator is commonly used.

These measures substantially improve the operating reliability of CU and the experience in exploiting the machine shows that with their regular execution failures arise in CU much less frequently than in the other units.

CHAPTER IV

Examples of Atypical Failures in M-20

Atypical failures are those which have, an infrequent, random character. In contradistinction to typical ones, atypical failures are more difficult to predict by preventive maintenance in view of the great variety and complexity of their effects in the various circuits of the machine. However, the experience accumulated in the analysis and elimination of such failures permits the time lost on their reappearance to be reduced. Part of the atypical failures may be reclassified as typical if we learn to reproduce the combinations in which they are expressed and to forestall them (for example, by the method of planned preventive maintenance), using logical and technical means.

Let us consider examples of atypical failures appearing in various units of the machine, the methods for detecting, and for eliminating them.

§1. Failures in CM

Example. After input of information a code cannot be read in MOZU.

We test the operation of CM by the instruction 0.20.0001.0000.0001 from the control console with the toggle Bl IR set to automatic mode. We write in register 0001 ⟨NVM-1⟩ = 7.77.7777.7777.7777 (the keyboard APU cleared). We obtain ⟨R1⟩ = ⟨NVM-1⟩ while ⟨R2⟩ = 0. There are no signals in the reception gates G1 CM at the points a_1, c_1, and there is no code reception signal at the points a_5, c_5 (VB KShCh). The element VB_{07-71} is incorrect; there is no output at a_7 for input signals to a_1 and a_0 of the element.

By this method it is possible to detect failures in the circuits generating the strobe signals St CS, St X, bit-position failures, etc.

Example. In the solution of a problem transient failures often occur in the machine (the tests run over wide limits). The failures have a random character; the instruction sequence of the program is disrupted.

The machine contacts are tapped. In tapping it is possible to observe contact failures on the front panel of the stack, in the connectors, poorly inserted lamps, bad solder joints, etc.

Example. Programs are often disrupted. At −42 V the test CM with repetition does not run; F_{zy} in the lower boards drop out.

The machine temperature is low.

Example. When the button "erase CM" is pushed the 32nd position in IR lights up.

To localize the site of failure we transfer from position to position elements in the read–write circuits of CM F_x, CS2, PF, etc. The failure is in one of these elements.

Example. Tests run over wide limits but programs are disrupted; most frequently standard programs exit to Aut. STOP.

To write their intermediate results the standard programs use the initial registers of CM. We test accessing of these registers at shifted nominal potential −45 V, +185 V.

Example. Certain positions in register address 1535 are not read. Oscilloscope analysis showed that the electronics operates normally and the failure is due to defects in the bit line.

To eliminate the defect we substitute the defective bit line by a spare line. Using Table 2, we find in the fourth graph the location of the auxiliary bit for Y = 15 (the 7th cartridge) and in the sixth graph the location of the defective bit line with address 1535 (11th cartridge). The input and output of the X-winding in the defective bit line are resoldered to the input and output of the X-winding of the auxiliary bit. The elimination of this failure is facilitated by the fact that the auxiliary bit for Y = 15 is free.

TABLE 2. Addresses of the Machine Stack

Cartridge No.	Auxiliary bits								Cartridge No.
	00—17	20—37	40—37	60—77	00—17	20—37	40—57	60—77	
1	01	31	21	11	10	20	30	00	32
2	03	33	23	13	12	22	32	02	31
3	05	35	25	15	14	24	34	04	30
4	07	37	27	17	16	26	36	06	29
5	11	01	31	21	20	30	00	10	28
6	13	03	33	23	22	32	02	12	27
7	15	05	35	25	24	34	04	14	26
8	17	07	37	27	26	36	06	16	25
9	21	11	01	31	30	00	10	20	24
10	23	13	03	33	32	02	12	22	23
11	25	15	05	35	34	04	14	24	22
12	27	17	07	37	36	06	16	26	21
13	31	21	11	01	00	10	20	30	20
14	33	23	13	03	02	12	22	32	19
15	35	25	15	05	04	14	24	34	18
16	37	27	17	07	06	16	26	36	17
17	41	71	61	51	50	60	70	40	16
18	43	73	63	53	52	62	72	42	15
19	45	75	65	55	54	64	74	44	14
20	47	77	67	57	56	66	76	46	13
21	51	41	71	61	60	70	40	50	12
22	53	43	73	63	62	72	42	52	11
23	55	45	75	65	64	74	44	54	10
24	57	47	77	67	66	76	46	56	9
25	61	51	41	71	70	40	50	60	8
26	63	53	43	72	72	52	52	62	7
27	65	55	45	73	74	44	54	64	6
28	67	57	47	77	76	46	56	66	5
29	71	61	51	41	40	50	60	70	4
30	73	63	53	43	42	52	62	72	3
31	75	65	55	45	44	54	64	74	2
32	77	67	57	47	46	56	66	76	1

Example. Failures have the same character as in the preceding example, with address 2531. The auxiliary bit is utilized.

From Table 2 we find the sixth graph the location of the defective bit line with address 2531 (15th cartridge). To eliminate the failure in this case it is necessary to remove the stack from the machine and carry out a number of repair operations whose purpose is to change the defective bit line for a free auxiliary bit with a different Y; we find a free auxiliary bit, disconnect it from its Y and assign it to the given Y.

Aside from defects in the bit line, other defects can arise in the stack influencing the operation of individual positions for different values of the register address.

Example. Halt occurs in the test CM with variable code. The register addresses vary, and the 14th bit is in error. Oscilloscope examination showed that the electronics operated normally and the failure is explained by a defect in the stack.

The failure is eliminated by substitution of the 14th bit by one of the spare (46-48) bits ("detour" of bit). Detour of a bit may be carried out by suitable resoldering at the connectors of the stack (the write and read connectors) or in the electronic circuits. In the first variant it is necessary to remove the stack from the machine. The second variant is simpler and takes less time. The 14th bit is replaced by the 46th, for which the input of element G1 (detour for writing) and the output of CS2 (detour for reading) of the 46th bit are soldered and the same circuits of the 14th bit are opened.

§2. Failures in AU

Example. In the execution of instruction "54" the signal "1" F "ω" is generated, while the contents of position 36 of R1 are equal to "1."

Cycling this instruction, we examine the cause of A "1" F "ω" (Sd, "1" F "0" in bit 45 of R1, P17, "1" 6 OR, "1" F "0" R1). The signals "1" F "0" R1 should not exist since there is a "1" in bit 36 of R1. $A2_{08-31}$ is not operating in the circuit "0 absent" of the 12 high-order bits of R1.

Example. In execution of the mantissa shift instruction on a zero code to the right by one bit with respect to address a "1" lights up in the 19th position of register R1.

Cycling this instruction we examine the cause of A "1" in the 19th position of R1. The DF in position 18 of R1 is oscillating.

Example. In the solution of problems with accessing to NVM-1 the program is lost after this access.

On execution of instruction 0.20.0001.0000.0000 the 13th bit in R1 is false.

Example. In execution of instruction "52" with respect to A3 instruction 0.50.0000. A2.0000 is blocked.

Cycling the instruction in automatic mode we examine DF in position 38 of EA. This DF sets to "1" on parasitic pulses.

Example. On execution of the instruction 0.07.NVM-1. NVM-2.0000 in the automatic mode transient failures occur on presence of "1" in positions 24, 26, 27, and 28 of NVM-1 and NVM-2.

We study the signals at the input to the carry look-ahead amplifier $A4_{07-49}$, since the error is expressed in a quartet of positions on execution of the given instruction. There is no output from c_6 of $A4_{07-49}$, which gives an appreciable total time delay of output signal with respect to the input.

Example. With repeated input of information to the machine with correct check sum the check sums differ in the high-order positions. The disagreement is unique.

Since the disagreement of the check sum with the known sum is unique and is expressed in the same positions (high-order), we assume that the error is occurring in the carry look-ahead circuits. To localize the failure we interchange $A4_{09-49}$ of the high orders (positions 33 to 36) with $A4_{08-49}$ of positions 29 to 32. The result is disagreement of the check sum in the other positions.

Example. Code is not fetched from NVM.

The instruction 0.00 NVM-1.0000.0000 is cycled (toggle Bl NVM set). There is no output from $c_8 A5_{07-18}$ to AU in the NVM output circuit.

Example. On execution of the instruction 0.04.NVM-1. NVM-2.0000 halt occurs due to the 3rd position of CLCO.

We interchange amplifiers $A2_{12-23}$ in the 3rd position of the counter of LCO with $A2_{12-22}$ of the 2nd position. On the next execution stop occurs because of the 2nd position.

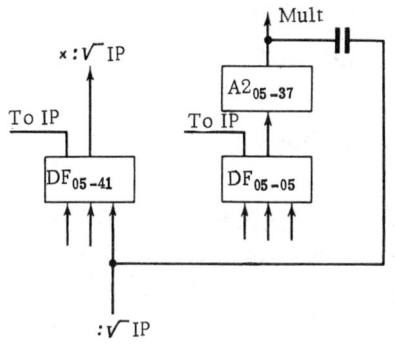

Fig. 9. Example of a failure not found from the signal lamps of CC.

Example. On execution of the operation "multiplication" the low-order product bits are incorrectly obtained in R1.

The circuit SdP2 for shift of the low-order bits in MA (1st position and AP MA) are examined in the 36th and 35th positions of R1. Flipflop DF_{01-13} (side a) for transmission of the code from AP MA to position 35 of R1 is not functioning.

Example. The division and square root operations are being disrupted. There are no parasites in accumulator and registers.

In inspecting the instruction in individual mode it was found that there is no transfer of carry "1" from position 4 to position 5 of CLCO. On interchange of DF "01" between the 4th and 3rd positions of CLCO the failure shifted to the 3rd position.

Example. The receiving flipflop RFRR does not reset. The RFRR circuit is investigated. A diode is short-circuited in the element $SB1_{10\ 51}$ of AU.

Example. In execution of the operation "+" the signals A "1" APMA and A "1" AP R1 are absent when the first address of R2 has any contents whatever.

In execution of this instruction we check the reason for the absence of these signals, using the oscilloscope. The transformer in A2 in the "0 absent" circuit of the first address of R2 is burned out.

§3. Failures in CU

Example. On execution of instruction 0.05.7771.7772.0000 there is no fetching of codes from R1 to R2. The "×" lamp is lit on the panel OCCC, but not the lamp "×: $\sqrt{IP}$." The operation ": $\sqrt{IP}$" is executed normally.

We examine the reason: the signals at the input-output of DF_{05-41} "×: $\sqrt{IP}$." There is no output from $A5_{05-37}$, although the multiplication flipflop DF_{05-05} is functioning (Fig. 9).

Example. On execution of the instruction 1.12.A1.A2.A3 there is no transfer to A2, although $RA < A1_{eff}$.

On execution of the instruction in the individual mode we note that the signal (+) No1 on P4 (transfer of $A1_{eff}$ to MA) is absent. A diode is short-circuited in the gate $G1_{03-45}$ in the network (+) No1.

Example. On execution of the zero instruction in the individual mode the k-th position of AA is not reset to "0" (is not reset even on general reset).

The failure is not examined in the automatic and cyclic modes, and therefore to localize the site of the failure we interchange AM and A2 in the corresponding positions. Either AM or A2 (side A) is being parasited in the k-th position of AA.

Example. With bias − 42 V it is not possible to start CSC from the button "start CSC" in the automatic mode: CSC operates as in the cyclic mode.

We examine the start CSC circuits with oscilloscope. The gate OV_{04-42} is not operating in the automatic start circuit of CSC.

Example. Halt occurs in all examples of test ω with generation of $\omega = 0$.

We study the generation of the tag $\omega = 0$ in some example of the test. F "0" ω DF$_{01-32}$ (side c) is being upset.

Example. In listing of the printing test only the "diagonal" lines of text are printed, beginning from $- + - 03.456789.01$, etc.

Instead of the contents of registers 0004-0031 only the contents of registers 0024-0031 is being printed; we therefore assume that the 5th position of AA is being excited. This error is not expressed in any other operations. DF$_{11-08}$ in the 5th position of AA is faulty.

Example. In operation "32" there is no transfer on CRA + 1.

In execution of instruction "32" in the individual mode it was detected that 16 of Table 3 is not being executed. In checking the signals at G1$_{07-26}$ the signal "1" PFSMA did not appear; the transformer in PFS2MA was burned out.

Example. The test CT is being disrupted. Control is not being transferred to the correct addresses.

On execution of one of the instructions in individual mode it is found that in standard formation of address the 4th position of address adder is being excited. AM or A2 in this position is incorrect.

Example. Address-modification errors are taking place.

On the test OT it was found that 3757 is being accessed in RA in place of 3777. The DF in the corresponding position is not functioning.

Example. On execution of OT "interrupt" is taking place on instruction "70" in place of "72."

In the execution of instruction "72" in the mode "instruction — suppress instruction" it was found that in place of "72" instruction "70" is sometimes selected. A2$_{08-04}$ of CU (in the 2nd position of OR) is not operating.

Example. Test CT is not being read from MD, test MM cannot be run (random errors in check sum on instruction "70"), frequent check-sum errors on input.

The signals in the circuit A "1" TK are examined in the "input" mode with toggles RKK, SP, OM set and AM reset. Synchronization is taken from P3. There is weak output from A2$_{06-44}$ in CU in the circuits A "1" F "K" and A "1" TTsS.

§4. Failures in the Input — Output Devices

Example. On data input check-sum errors occur. Information is incorrectly stored in CM. On repeated input the check sum obtained is different each time.

With toggles RKK, SP, OM set and toggle AM reset we start input. The character of the parasitic signal is clearly visible on CC. The monostable MS$_{09-47}$ in the circuit for starting CSC from MM is being upset.

Example. After data input the deck of cards in the stacker is disordered.

The pair of stacking arms in CR is not operating.

Example. In putting out lines of test print at bias -38 V for the basic nominal potential print is "disrupted" on zeros.

We examine the signals of the print counter ("512"). Synchronization is taken from start of print or the first SPD. Parasitic signals appear in $A2_{02-23}$ (the 1st position of the counter is normal, the 2nd "floats." The output of $A2_{02-23}$ on c_9 is noisy).

Example. In output of differing information for printing the same contents are printed all the time.

The circuit for clearing R1MMCU is examined. Synchronization is taken from start of print. The transformer in $A2_{02-56}$ in the pulse train GI "00" (output c_7) has burned out and the tube in element $G1_{02-56}$ (side a) in the R1 clear circuit has failed, etc.

Example. In output of data to printer zeros are printed in the three low-order positions.

On output of data the strobing signal is examined in the three low-order positions of register MMCU. Side a of amplifier $A5_{05-63}$ is inoperative (no output from a_8), removing strobe from the three low-order positions of RCh MMCU.

Example. In output of a test to printer the printer prints zeros continuously. On blocking of print clear on CC output is normal.

Constant clearing of print is taking place. Element S_{02-58} in Pr clear is faulty.

Example. On output to printer the printer prints zeros continuously. On blocking print clear on CC the result is not changed.

We investigate the signals "3072" arriving at the print control circuit. The control signals "3072" are absent from WrPr, C, etc.; failure is in amplifier $A5_{05-67}$ (side a).

Example. On output of large arrays of numbers certain numbers in the arrays are incorrectly printed. The failures are infrequent.

We check the voltage applied to extinction of the thyratron register (it seems that the thyratron register is sometimes not being cleared). The grid resistor in one of the KP2 elements has burned out.

Example. The top one or two lines of print have dropped out. The repetition frequency is once or twice per 10 listings. The failures could not be found in the electronics.

We check all nominal potentials in the high-speed printer blocks. The voltage $U = 200$ V on Pr is low.

Example. In output of lines of "test print" the ones in the second group (... 010101) are not printed.

A Rep 3 is not operating in the 5th position.

Example. On change of bias -40 to -45 V the tube "start M" burns out.

The monostable MS in the circuit "start M_{erase}" is excited, as is easily observed at the point a_3 of the element MS.

In this chapter we have limited our consideration to only a small number of examples, characterizing possible failures and transient failures of the general-purpose computer M-20. The systematization and study of the causes of their appearance permit the time to find and eliminate them to be substantially reduced. The method for locating transient failures and failures may be used not only in the exploitation of a machine of the type M-20, but for others.

CHAPTER V

Certain Questions of Improving the Machine During Its Exploitation

To improve the operating reliability of the machine and extend its possibilities the following basic measures are applied.

1. Reinforcement of weak and most heavily loaded circuits and elimination of oscillations.

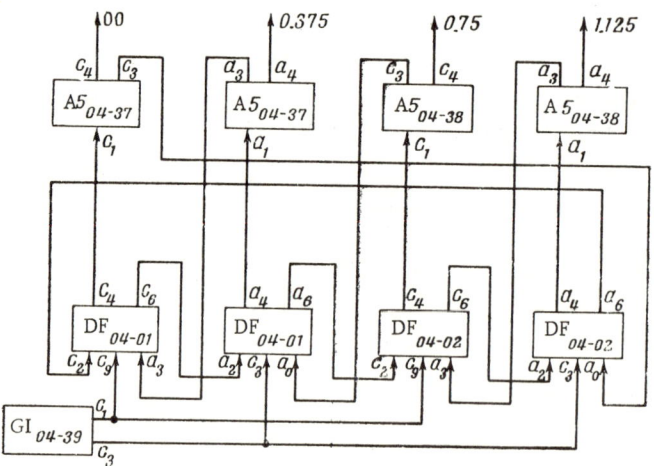

Fig. 10. Circuit changes in the pulse train block.

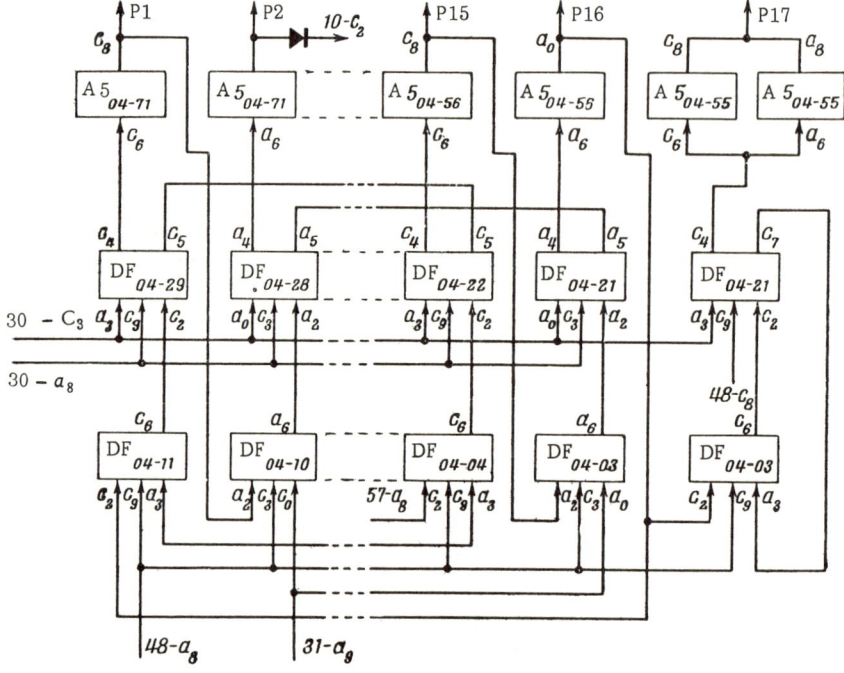

Fig. 11. Changes in CSC circuit.

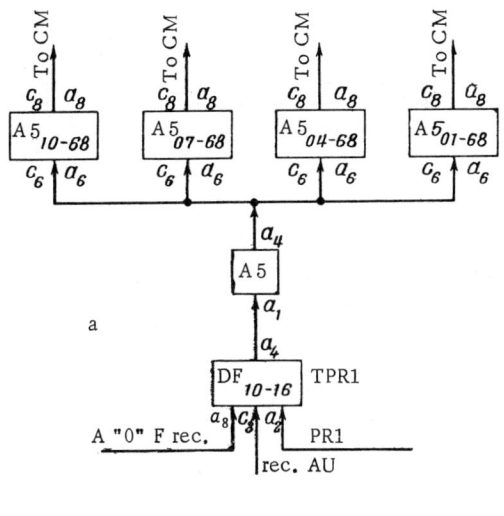

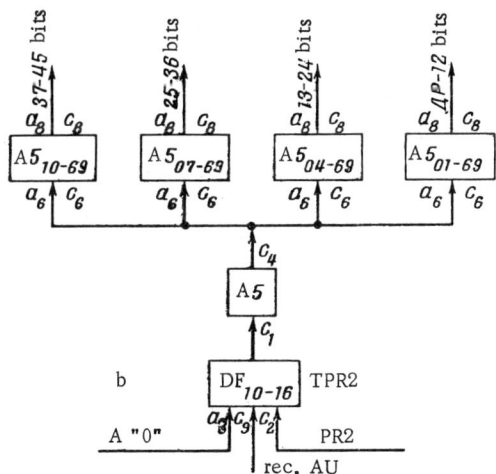

Fig. 12. Circuit changes in the reception circuits. a) In R1; b) in R2.

2. Introduction of auxiliary circuits to localize the points of intermittent failures (Chapter II, §2).

3. Reduction of the time for execution of certain operations to acclerate program execution in the processing of information from a physical experiment.

§1. Reinforcement of Weak and Most Heavily Loaded Circuits and Suppression of Oscillations

The following measures were taken in M-20 to reinforce the weak circuits and suppress oscillations.

1. The amplifiers A2, subject to regeneration, were eliminated from the synchronization block (Fig. 10).

2. For stable operation of CSC operation the setting of the matching flipflops to "1" is taken directly from the outputs of amplifiers A5, and not from the flipflop outputs (Fig. 11).

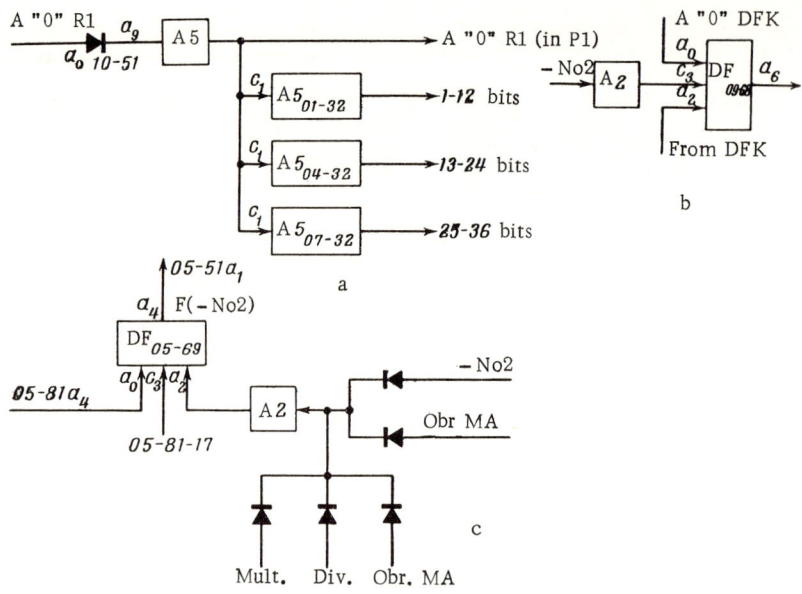

Fig. 13. Changes in the circuits. a) A "0" R1; b) F "DK" MA; and c) A "1" DF (−No2).

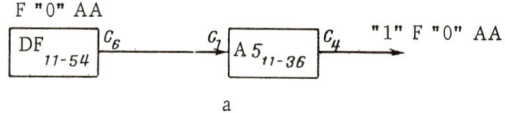

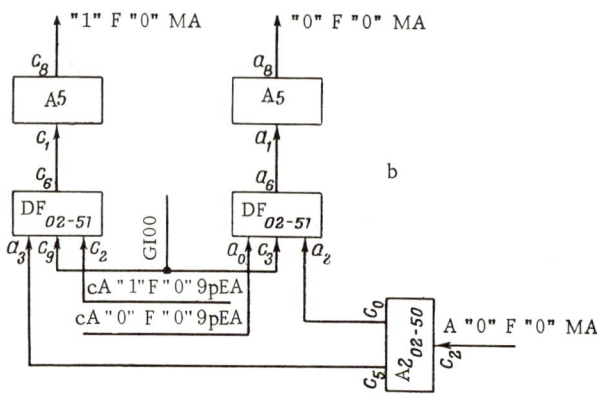

Fig. 14. Changes in circuits. a) F "0" AA and b) the paraphase trigger F "0" MA.

3. An additional element 5 is connected in the reception circuits of registers R1 and R2 (Fig. 12a, b).

4. An amplifier A5 is connected in the circuit A "0" R1 (AU block 2) after the $OR1_{10-51}$ gate (Fig. 13a).

5. The pulse train DFK MA is reinforced by means of the amplifier A2 (Fig. 13b) and the circuit for setting DF_{05-64} (−No2) to "1" (Fig. 13c).

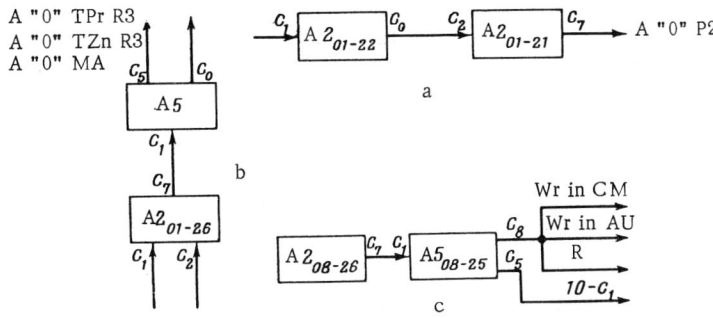

Fig. 15. Changes in the circuits. a) A "0" P2; b) the AU reset; and c) CCO CM.

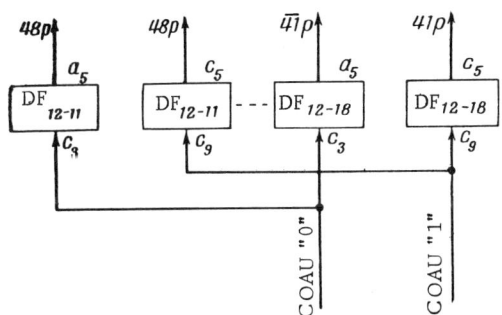

Fig. 16. Decoupling of circuits COAU "0" and COAU "1."

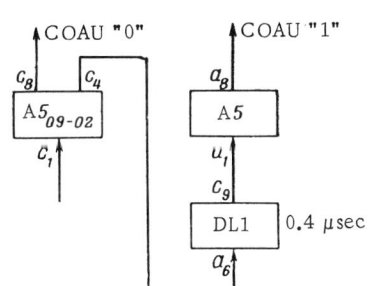

Fig. 17. Introduction of delay into the circuit COAU "1."

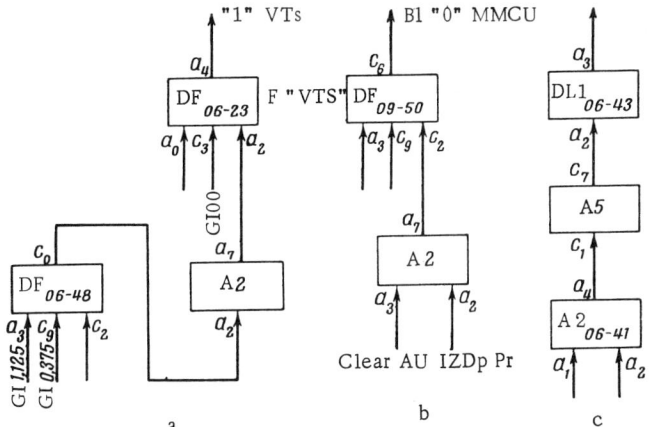

Fig. 18. Changes in the circuits. a) A "1" F "VTs," A "1" FB1 "0" MMCU, and the matching flipflop NC.

6. An amplifier A2 is connected into the circuit for setting the normalization flipflop to "1" in the circuit "LCOSd," and the amplifier A2 operating in the train "00" is removed for the same reason as in 1.

7. The output signal F "0" AA is reinforced by means of an amplifier A5 (Fig. 14a).

8. The outputs of the paraphase flipflop F "0" MA are reinforced by insertion of elements A5 (Fig. 14b).

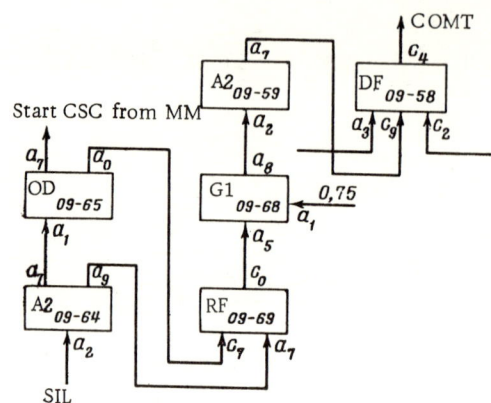

Fig. 19. Changes in circuit for generating the signal COMT.

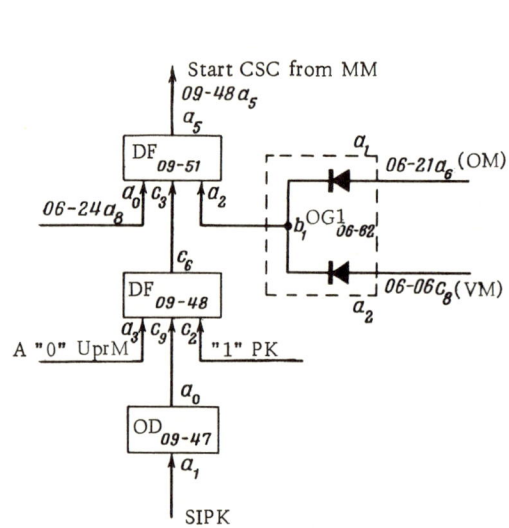

Fig. 20. Changes in the circuit start CSC on input from punched card.

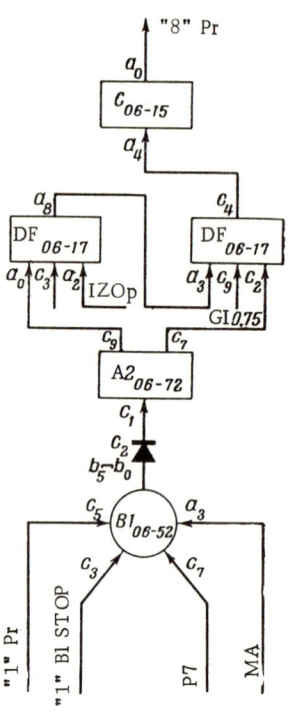

Fig. 21. Circuit for control of octal print.

9. The signal A "0" P2 in addition (Fig. 15a) and in the circuit for clearing AU (Fig. 15b) is reinforced.

10. In the block CCO CM the signals R and Wr CM have been reinforced (Fig. 15c).

11. The networks COAU "0" and COAU "1" are decoupled (Figs. 16 and 17).

12. The circuit for setting to "1" F "VTs" (Fig. 18a), the flipflop Bl "0" MMCU (Fig. 18b) and the matching flipflop, setting to "1" T "NC" (Fig. 18c) are strengthened.

13. The circuit for generating the strobing signal COMT is changed (Fig. 19).

14. The circuit for starting CSC on punch card input is changed (Fig. 20).

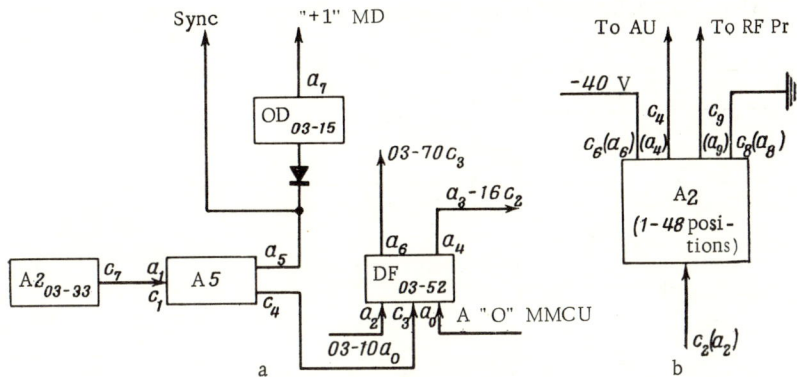

Fig. 22. Changes in the circuits: a) generating the signal "+1" of MD and b) the output networks of the positional amplifier elements A2 of CM.

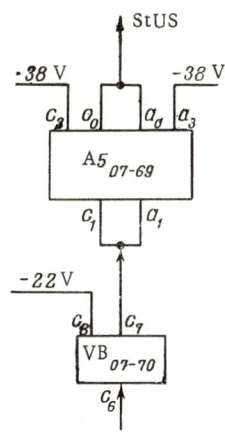

Fig. 23. Changes in the circuit for generating the signal StUS.

15. A circuit is constructed for operation of octal printout by CN-0500 (Fig. 21).

16. For reliable triggering of the monostable in the circuit for generating the signal "+1" MD is not taken from an element A2, but from an additional element A5 (Fig. 22a).

17. The output winding of the element A2 in the CM reading circuits have been interchanged (Fig. 22b).

18. The differentiating network at the input to A5 has been eliminated from the circuit for generating the signal StUS (Fig. 23).

To improve the operating reliability of the individual blocks and devices of the machine other changes have also been made in the basic circuits of the machine (MT tracks placed in parallel, decoupling of the circuits +300 V for MMCU and Pr, modification of the circuit "0 absent" in AU, reinforcement of the reception circuit on OR, etc.). The complex of measures carried out has permitted the uptime of the machine to be increased by 40%.

§2. Reduction of the Execution Time of Certain Operations

All changes in the execution of any operation are registered in Table 3. Let us consider the more fundamental ones.

Operations of "Addition, Subtraction, Subtraction of Absolute Values"

Let us consider the two most interesting cases.

1. To normalize unnormalized numbers in the processing of data from a physical experiment programmers widely use the method of addition to zero of the unnormalized number. However, an addition to zero unnecessary operations of equalization of exponent occur before normalization. The higher the order of the unnormalized number, the longer the time expended on the execution of this operation; the maximum difference of order is $177_8 = 127_{10}$, which takes $127 \times 1.5 = 190.5$ μsec of lost time. On the average 95 μsec are lost. To block equalization of exponents on addition to zero, an additional signal "0" F "ω" is introduced into the circuit for starting LCO SdPSmCH or SdPR1 (CU, block 07) (Fig. 24). Since in this case, when one of the numbers is equal to zero, a standard setting to "1" F "ω" takes place, the unnecessary shifts in the accumulator or register R1 do not take place.

TABLE 3. Changes in the Tables of Operation Execution

No. in M-20 operation table	Was			Instructions	No. in M-20 operation table	Became		
	Microoperation	Conditions				Microoperation	Conditions	
		1	2				1	2
				Addition, subtraction, subtraction of abs. values	17a	+No1	+− | − | "1" F "ω"	12
19	Start LCO Sd PMA, A "0" P2, A "1" F LCO, +No1	+− | + | "1" RFSEA	12		19	Start LCO Sd PMA, A "0" P2 A "1" F LCO, ++No1	+− | − | "1" RFSEA "0" F "ω"	12
20	A "0" P1, A "1" F LCO +No1	+− | −"0" F "−0" EA	12		20	A "0" P1, A "1" F LCO, ++No1	+− | − | "0" F "−0" EA, "0" F "ω"	12
23	Start CSC	From LCO Sd			23	Start CSC	From LCO Sd output pulse LC	
1	A "−65" EA	Kh	4	Multiplication	1	A "+64" EA Obr EA	Kh	4
1	A "0" F "ω"	Kh (b)	3	Output of lowest-order bits of product	1	A "0" F "ω"	Kh (b)	4
1	A "0" F "ω"	IP, Sd	3	Change of exponent, shift	1	A "0" F "ω"	Sd	4
3	A "−24" EA	IP, Sd "0" 5OR	4		3	PAP1	IP, Sd "0" 5 OR	4
4	Start LCO Sd	A "−24" EA	(4)		4	Operations completely excluded		
5	Stop CSC	A "−24" EA	(4)		5			
6	Start CSC	From LCO Sd			6			
8	A "−65" EA	IP, Sd	6		8	A "+65" EA	IP, Sd	4
24	A "0" EA	Sd	13		24	Obr EA		13
8	PR2	LOp "1" 6OR	11	Logical operations	8	A "0" EA A "0" MA Excluded	Sd	
				Write code in CM	11a	(+) R1	LOp "1" OR 00	13
					4a	(+) R2 PR2	"51" OR 00	9
5	VR1	00	17		5	VP1	"0" 6 OR	17

To transmit an unnormalized number to accumulator from register R1 when the second number is equal to zero a signal +No1 (+ − \| − \|, "1" F "ω" P12) is generated. Transmission of the second number from register R2 to accumulator on P10 (+ − \| − \|, P10, +No2) and further normalization of the number take place in the standard way.

2. In execution of the operation "equalization of exponents," if the order of one of the numbers is too great and the other too small (so that the difference between them is greater than 37_{10}), one of the numbers will be pushed out beyond the limits of the register. Further equalization of exponents is useless. One possible variant of realization of the circuit blocking equalization of the exponents when the difference between them exceeds 37_{10} is shown in Fig. 25.

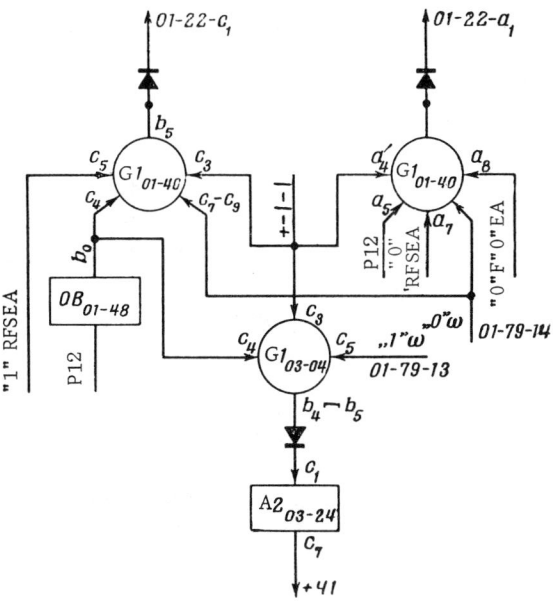

Fig. 24. Changes in the circuit start LCO SdPSmCh and SdPR1.

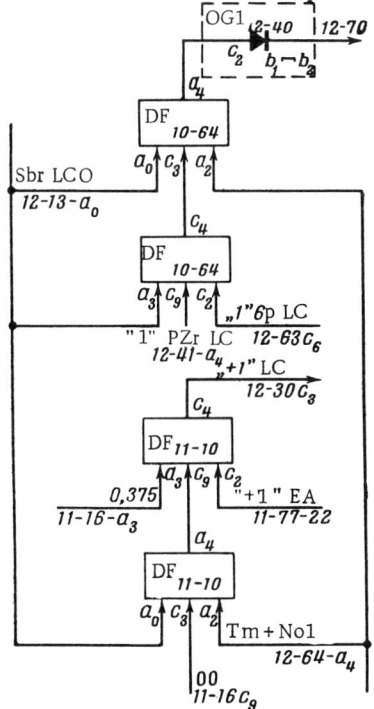

Fig. 25. Circuit for blocking equalization of exponents for differences greater than 37.

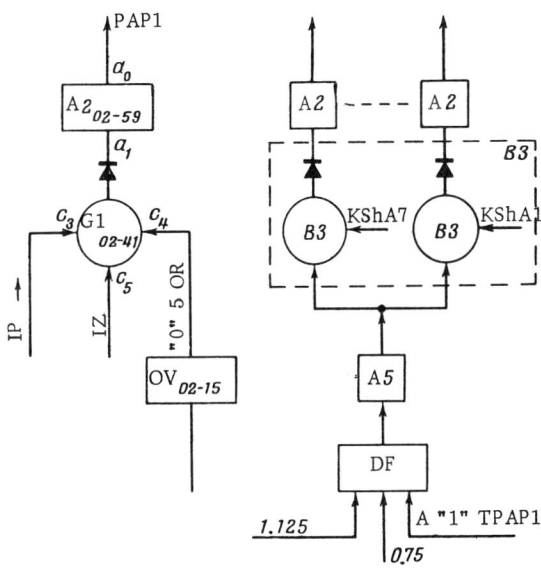

Fig. 26. Changes in the circuit for execution of "shift" and "modification of exponents."

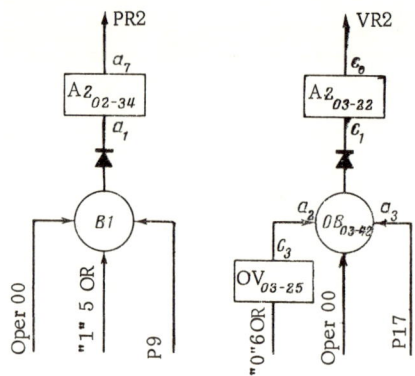

Fig. 27. a) Circuit for generating signals A "+64" and Obr. EA; b) changes in the circuit "logical operations."

To count the number of shifts executed counter LCO is used. At each shift addition "+1" EA takes place, which is used for A "1" DF_{11-10}, generating "+1" LC. The criterion for terminating shift is "1" in position 6 of CLCO and carry "1" from the 3rd position of counter LCO ($100101 = 45_8 = 37_{10}$), i.e., the number of shifts is increased by two (39 shifts take place). Further start, generation of the signal "+No1," clearing of counter LCO are executed in the standard manner; the output of flipflop DF_{10-64} over $OG1_{12-46}$ sets to "1" TU "0" M, extinguishing the counter LCO and preparing the circuits "start CSC" and +No1. The maximum possible difference of exponents of two numbers is equal to $177_8 = 127_{10}$.

The maximum time economy is $(127-39) \cdot 1.5 = 132$ μsec. The mean time is about 66 μsec.

Operations of "Change Exponents and Shift"

To reduce the time for executing the operations "change of exponents and shift" circuits have been proposed, composed of standard elements (Fig. 26), and the corresponding changes are listed in Table 3. The number of necessary shifts with respect to address is defined by direct reception of the address in E1, which permits the operations "−24" EA, Start LCO, SdL1R1, Stop, and Start CSC to be eliminated. The operation execution time is substantially reduced (by about 36 μsec), since 24 shifts drop out for a constant from the positions of the second address register R1 into the exponent positions E1. The circuit of amplifier A "24" EA is used directly to generate the signal PAP1, and to reduce the exponent to a natural value in place of A "−65" EA, A "64" EA and Obr EA by P14 are introduced.

In the operation "multiplication" A "−65" EA is realized in a similar way (cf. Table 3). See Fig. 27a for the formation of the signals A "+64" EA and ObrEA.

Logical Operations

Negligible changes have been made in the logical operations. The transmission of information from register to register (VR1 and PR2 on pulse P11) is eliminated from the operations of "logical product" (operation "55") and "logical sum" (operation "75"). To form the logical sum xVy (Table 3) signals (+)R1 and (+)R2 are generated on the 13th pulse CSC (Fig. 27b). The codes from registers R1 and R2 are applied to the same accumulator input, where the sum is formed.

§3. Other Changes in the Circuits of M-20

Certain changes in the instructions of M-20 and in the logic of execution of certain of its operations have been realized in application of the machine in a physical experiment. These changes have been published in various articles [1-6], or are in press in other editions.

The development of methods for using computers in the conditions of a complex physical institute permits the time for putting new machines of type M-20 into operation but also of various ancillary devices and installations for servicing a physical experiment, to be substantially reduced.

In conclusion we may point out further directions for increasing the machine uptime.

1. Constant study of the reasons for machine failures, systematization of the accumulated experience, and its exchange among the various organizations concerned with questions of exploitation of the machines.

2. Improvement of the machines, directed towards increasing the operating reliability of its elements, units, and devices.

3. Improvement of the methods of preventive maintenance and testing, broadening of the function of circuit checking and the possibilities of the control console; introduction of automatic monitoring and program methods of locating failures.

4. Application of high-quality electronic components.

5. Improvement of the qualifications of the staff servicing the machine, etc.

LITERATURE CITED

1. A. V. Kutsenko, Yu. V. Stupin, and I. V. Shtranikh, FIAN Preprint No. 82, Moscow (1966).
2. V. V. Gavrilov, A. V. Kutsenko, Yu. V. Stupin, and B. V. Subbotin, Pribory i Tekh. Eksp., No. 2 (1967).
3. V. V. Gavrilov, A. M. Klabukov, A. V. Kutsenko, Yu. V. Stupin, and I. V. Shtranikh, FIAN Preprint No. 132, Moscow (1966).
4. V. V. Gavrilov, B. G. Minaev, and Yu. V. Stupin, FIAN Preprint No. 40, Moscow (1966).
5. V. V. Gavrilov, B. G. Minaev, and Yu. V. Stupin, in: Digital Computers and Programming, Vol. 2 [in Russian] (1967).
6. Bulletins of KEVM, Nos. 1-3, Moscow, VTs AN SSSR (1961-1965).

Appendix

Glossary of Acronyms and Abbreviations

"Standard" translations have been found and introduced for the following acronyms and abbreviations.

Abbreviation used	Transliteration of abbreviation used in Russian text	Meaning
A1	A1	First address of instruction
A2	A2	Second address of instruction
A3	A3	Third address of instruction
AA	SmA	Address adder
ACC	PAU	Autonomous control console
Ad	Sm	Adder
AM	BS	Adder module
AM	VM	Auxiliary marker
A*	Y*	Amplifier*
AP	DR	Additional position
AR	US	Reading amplifier
ARep	UV	Reproduction amplifier
AU	AU	Arithmetic unit
AUA	BSAU	Arithmetic unit accumulator
Aut. STOP	Avost	Automatic stop

Abbreviation used	Transliteration of abbreviation used in Russian text	Meaning
AWr	UZ	Writing amplifier
BPC	Blk	Block parity check
C	Sch	Counter
CC	PU	Control console
CCO	TsUOp	Central control of operation
CLCO	SchMUOp	LCO counter
CM	BP	Carry module
CM	MOZU	Core memory
CN	UCh	Conditional number
COAU	VKAU	Code output from AU
COMT	VKML	Code output from MT
CR	ChU	Card reader
CRA	KRA	Command address register
CS	US	Control signal
CSC	TsUS	Central signal control
CT	KT	Complex test
CU	UU	Control unit
DF	TD	Dynamic flipflop
Div	Del	Division
DL	LZ	Delay line
E	P	Exponent
EA	SmP	Exponential adder
F	T	Flipflop
G	VL	Gate
G*	V*	Gate*
IR	RK	Instruction register
IS	IS	Interpretive system
LC	SchM	Local counter
LCO	MUOp	Local control of operation
MA	SmCh	Mantissa adder
MD	MB	Magnetic drum
MM	MZU	Magnetic memory
MMCU	UUN	Magnetic memory control unit
MS	OD	Monostable
MT	ML	Magnetic tape
NC	NTs	Normal cycle
No	Ch	Number (addend, augend, etc.)
NVM	DZU	Nonvolatile memory (word organized toggle banks)
OC	KOp	Operation commutator
OG	SB	OR gate
OR	ROp	Operation register
P*	I*	Pulse*
P	Pf	Punch
PC	SchPch	Print counter
PF	TP	Paraphrase flipflop
Pr	Pch	Printer
R	Cht	Reading
R1	R1	First arithmetic register

Abbreviation used	Transliteration of abbreviation used in Russian text	Meaning
R2	R2	Second arithmetic register
RA	RA	Address register
RF	TP	Receiving flipflop
RR	RR	Result register
S...	Zn...	Sign...
SP	SI	Synchronization pulse
SPD	SIB	Synchronization pulse drum
T	Tb	Toggle
Wr	Zr	Write

_____
* Followed by a number

No "standard" translation could be located for the following abbreviations. They have therefore been rendered as transliterations of the forms found in the Russian text.

APU	IZOp	KShCh	OD	PK	RCh	SdPSmCh	TK	TTS	VB
Bl	K	LOp	OM	POM	RKK	SdTs	TLPR	TTSR	VL
DK	Kh	M	Oop	PR	S	SIL	Tm	TTsS	VM
F_x, F_z	KK	Ma	OT	PRR	Sd	SIPK	TPAP	TU	VR
GI	KP	NPCh	OV	PsmCh	SdL	StKH	TPr	TZn	VTs
IP	KPch	OB	PAP	PTs	SdP	StUS	TsS	UprM	VTsK
IZ	KShA	Obr	PB	PZr	SdPR	TG	TsV	VAPU	

SOME PROBLEMS IN ANALYZING THE DYNAMIC STRUCTURE OF AN OBJECT FROM THE STEADY-STATE SIGNAL

L. I. Gudzenko

§1. Inverse Analysis Problems

Because of the increased interest in studying more and more complex effects, a greater amount of attention is being given to experimental design as well as an analysis of the results of passive observations. It is not always suitable to use the traditional natural-science method of study, i.e., the sequence of operations which include the obtaining of experimental data (uncovering the qualitative characteristics of the effect and constructing a model, writing equations for the model, and solving these equations), comparing the solutions with experiments, refining the model, etc. In the field of cybernetics, a "black box" scheme has been proposed for studying complex objects when the compiling of the measurement results is already a very common problem; in this method the actions performed by the experimentor on the object are described formally as a set of excitations or perturbations. Proceeding from the regular comparison of the responses of the object ("at the box output") with the excitations ("at its input"), this scheme is characterized by a somewhat different sequence of operations: the choice of an *a priori* class of equations for the model; comparison of the responses with the excitations and obtaining specific equations for the models; the construction of the model. In the simplest variant the relaxation toward stable limit motion as a result of an initial deflection is analyzed. In principle, the problem of finding equations for the model is solvable if the excitations at the input of the black box are fairly complete and all of the degrees of freedom of the object which are of importance in the observed effect can be obtained from the responses of the object.

We shall illustrate this by an elementary example. Let the signal $U_0(t)$ of an autonomous object be described by an ordinary linear differential equation with stable stationary point $U_0 \equiv 0$:

$$\frac{d^q U_0}{dt^q}(t) + \sum_{m=0}^{q-1} a_m \frac{d^m U_0}{dt^m}(t) = 0. \tag{1}$$

We let $\{U_0^{(m)}\}$ represent the initial conditions $\frac{d^m U_0}{dt^m}(0) = U_0^m$, $m = 0, 1, 2, \ldots, q-1$. Assuming for simplicity that Eq. (1) does not have multiple characteristic numbers we obtain

$$U_0(t) = \sum_{l=1}^{q} C_l \exp(k_l t), \quad t \leqslant 0. \tag{2}$$

The characteristic equation $k^q + \sum_{l=0}^{q-1} a_l k^l = 0$ establishes a mutually single-valued relationship between the set of roots $\{k_l\}$ and the vector $(a_0, a_1, ..., a_{q-1})$. On the other hand, for a given choise of characteristic numbers we have q equations $\sum_{l=1}^{q} C_l (k_l)^m = U_0^{(m)}$ $(m = 0, 1, ..., q-1)$, which uniquely relate the vectors $(c_1, c_2, ..., c_q)$ and $(U_0^{(0)}, U_0^{(1)}, ..., U^{q-1})$. Clearly, each specific description of the signal $U_0(t)$ makes it possible to find only those values of k_l for which (in accordance with the initial conditions $\{U_0^{(m)}\}$) the coefficients C_l prove to be nonzero. Here, the purpose of the analysis is to calculate the coefficients $\{a_l\}$ from the recording of $U_0(t)$; therefore we can assume that each of the vectors $(U_0^{(0)}, U_0^{(1)}, ..., U_0^{q-1})$, for which all $C_l \neq 0$ makes it possible to study the object without additional excitations, i.e., one such initial deflection realizes a complete set of excitations for the object.

This discussion loses its meaning if the experimentor is forced to draw conclusions by observing the steady-state motion of this object because he is not able to estimate the state of the autonomous object. There is a fairly broad region of application of this type of problem, i.e., the number of objects which cannot be regulated is greater than would appear at first glance. Typical astrophysical objects such as stars, galaxies and quasars fall into this category in which man is not able to noticeably affect the processes taking place. In addition, many medical and industrial objects for which non-normal operation is not desirable belong to this category. Here, naturally, we refer to problems involved in analyzing the results of recorded nonrepeatable observations. Moreover, in the initial stage of the study it is convenient to convenient to consider an object "unregulatable" if it is quite accessible to the experimentor, fairly complex, and its interaction with the environment, although being unimportant when studying a given effect, can strongly complicate the analysis.* In addition, in the past, it was not quite clear just how accessible was the set of excitations used by the experimentor.

When studying an unregulatable object it is of great importance to know that the recorded signals do not always describe in sufficient detail the behavior of the object as a whole but carry "errors" in the individual motion of a large number of its separate microparts of the same type. The simplest mathematical model for the formation of a signal U(t) in a time t which allows for this can be given by an equation of the form $D_0[U] = F(t)$, where F(t) are small short-correlation Gauss fluctuations for the exciting steady-state dynamic change in the signal which is produced according to the equations $D_0[U] = 0$. The class $\{D_\alpha\}$ of operators generating the dynamic equations $\{D_\alpha\}[U] = 0$ and, as a particular case, the equation $D_0[U] = 0$, which dynamically approximates the observed manifestations of the analyzed object, is determined from *a priori* information on the object, from similar criteria (allowing for the properties of objects familiar to the researcher and, from his point of view, similar to the one studied) and also from the principle of simplicity of description, expressed according to the ideas of the researcher as to the essence of the problem and its technological level.

The black-box scheme recommends that after the class $\{D_\alpha\}$ of dynamic operators is obtained we should seek (by numerically processing the signal U) the element D_0 in this class which produces the dynamic equation of the object to be studied. Then, from the form of D_0, it is necessary to use familiar natural laws to construct the physical, chemical, biological economic (whichever corresponds to the problem) model, rather than the mathematical one; this type of model will be known as a "natural" model. Here, most of the difficulty must un-

---

*Remaining within the framework of the idealized scheme of analysis described in this model, not allowing for the effect of internal fluctuations in the object, linearity, its external noise, etc.

avoidably appear as a result of the considerable effect on the observed phenomenon of laws of nature completely unknown to us.

In the first stage of analysis – isolating the class $\{D_\alpha\}$ – we are faced with common problems: from *a priori* data on the studied object it is first necessary to chose one of the various types of operators, i.e., functional operators, ordinary differential operators, differential operators with delayed arguments, partial differential operators, integral operators, or integral-differential operators. It then follows that we must know as much as possible the very broad class of the chosen type of operator. By allowing for more detailed *a priori* information we restrict ourselves to a consideration of linear or weakly nonlinear operators with constant coefficients. In this type of consideration, it is evidently understandable to allow not only for the a priori properties of the analyzable object but also preliminary information obtained from the signal. It would be interesting in this respect to know what laws for signal behavior would make it possible to ignore any given class of operators as not explicitly corresponding to the observed manifestations of an object. However, because of the absence of final complete results, a discussion of these problems which is associated with the first stage of analysis would actually consist of a rehashing of several of the previously considered particular problems. An even more independent approach is to immediately study each new object at the final stage, i.e., construct a "natural" model. Here, as yet, not even isolated problems have been commented on and the resultant difficulties and methods for overcoming these difficulties could be illustrated by several models constructed as specific examples.

This article is devoted to a discussion of the statement of the problem and method for solving this problem in the simplest variant of the second stage involved in analyzing the structure of an unregulatable object, i.e., the isolation of the operator D_0 corresponding to the statistical properties of the recorded signal from the class of dynamic operators $\{D_\alpha\}$, assuming this class is previously known [1-4].

The development of methods for seeking a solution to an equation with a given function from a particular class refers to "inverse" analysis problems. The name "inverse," which seems to imply that such problems are secondary should be assumed to be conditional; it is determined by the history of the development of exact sciences. It has already repeatedly been noted that in the fundamental natural-science problems the researcher is directly concerned with inverse problems when the motion of an object is observed and it is necessary to understand its mechanism. "Direct" methods, i.e., finding the motion of an object of known structure under specified conditions usually has a narrower (or technical) area of direct application in science. However, information about the solutions of varous direct problems in analysis is often used in inverse schemes in order to understand the structure of the object from its behavior; here, the trial and error method is used to obtain the equations. However, such a method of study is not always effective or economical. The usual estimates as to the relative complexity of different descriptions of an effect, considerations as to the reliability of the conclusions, etc., apply to the traditional direct method of natural-science studies. When successively analyzing a given effect within the framework of the inverse method these representations often prove to be unjustified. The fact that there is as yet a small number of natural-science examples which are considered using the black-box technique without an input indicates that the difficulties in the direct and inverse schemes do not occur at the same points. For example, the basic step in the direct scheme, that of finding solutions to the equations of the object, is actually absent in the inverse scheme since in this case the researcher is first involved with the signal, i.e., with the prepared solution to the equations of the object.

§2. Statement of the Problem

Let a signal $-\Theta/2 \leq t < \Theta/2$ be recorded over an observation interval $U(t; x)$ for different parameters $x_i \in \boldsymbol{x}_i$ ($i = 1, 2, \ldots, Q$), associated by the notation $\mathbf{x} \equiv (x_1, x_2, \ldots, x_Q)$. Here we shall assume that there is a dynamic relationship between the signal and the parameters $\mathbf{x}$. A very different "natural" meaning can be obtained for the real quantities x_i by considering them together or separately. The values of x_i can express the dimensions of an object, the concentration of chemical reagents, flow rate in blood vessels, the characteristics of a branch of economics, etc. Finally, the values of x_i can also constitute simply a set of prime natural numbers; this leads to the simplest variant for a multidimensional signal:

$$U(t; \mathbf{x}) \sim (U_1(t), U_2(t), \ldots, U_Q(t)).$$

Hencefore, the isolation of the time t (or the phase analog) from all $(Q+1)$ arguments of the signal is associated with the subsequent importance of obtaining all possible dependences on time; according to this, the cause has smaller values of t than its effect. As is not difficult to understand, it is not sufficient to analyze the structure of an autonomous object from its steady state signal allowing only for the dynamic dependence on time: even a detailed recording of the unique steady-state solution to dynamic equations $D_0[U; \mathbf{x}] = 0$ does not make it possible to obtain the structure of operator D_0 under actual conditions. Thus, from the position of the stationary point $U_0 \equiv 0$ of the simplest ordinary differential Eq. (1), not only is it impossible to estimate the numerical values of the coefficients a_l but it is also impossible to obtain the order of the equation. This drawback to the usual black-box scheme for studying an "unregulatable" object from its steady-state signal also makes it necessary to allow for internal fluctuations in the equations for the signal of the object.

We shall limit ourselves to a consideration of autonomous objects whose dynamic equations.

$$D_0[U(t; \mathbf{x}); \mathbf{x}] = 0 \tag{3}$$

do not explicitly contain the time t. We assume D_0 is a real differential operator of form

$$D_0[U; \mathbf{x}] \equiv \frac{d^q U}{dt^q} + a\left(U, \frac{dU}{dt}, \ldots, \frac{d^{q-1} U}{dt^{q-1}}; \mathbf{x}\right). \tag{4}$$

For each fixed values of the parameters $\mathbf{x}$, the variables $U, \frac{dU}{dt}, \ldots, \frac{d^{q-1}U}{dt^{q-1}}$ are treated as the coordinates of the image points U of the object in phase space $R_q(\mathbf{x})$ of its dynamic system. In addition to any solution $U_0(t; \mathbf{x})$ of autonomous dynamic Eq. (3), the function $U_0(t; +\tau; \mathbf{x})$ will also be a solution for any τ. In the space $R_q(\mathbf{x})$, the set of points $\{U_0(t+\tau; \mathbf{x})\}$ forms a unique geometrical image corresponding to the solutions for different values, i.e., the dynamic trajectory $\mathbf{U}_0(\mathbf{x})$. Assume that in region $G(\mathbf{x})$ the conditions of the Theorem for the existence and uniqueness of a solution to differential Eq. (3) are satisfied for each $\mathbf{x}$ $(x_i \in X_i, i = 1, 2, \ldots, Q)$ so that one and only one dynamic trajectory passes through each point $\mathbf{U}$ of region $G(\mathbf{x})$. If the trajectory $\mathbf{U}_0(\mathbf{x})$ does not degenerate into a stationary point, in a sufficiently small neighborhood $G(\mathbf{U}(\mathbf{x}))$, $G(\mathbf{U}_0(\mathbf{x})) \subseteq G(\mathbf{x})$, in which the orthogonal projections of point $\mathbf{U}$ on this trajectory are uniquely defined, we can introduce the phase $t'(\mathbf{U})$ of the image point $\mathbf{U}$ associated with the motion of $\mathbf{U}_0(t, \mathbf{x})$. From the definition of phase, $t'(\mathbf{U})$ is equal to that value of t which, in accordance with the law $\mathbf{U}_0(t, \mathbf{x})$, results when point U is projected onto the trajectory $\mathbf{U}_0(\mathbf{x})$. By studying the behavior of the object near a nondegenerating (into a point) trajectory $\mathbf{U}_0(\mathbf{x})$ we can then convert in the signal $U(t; \mathbf{x})$ from the argument t to the phase $t' \equiv t'(\mathbf{U}(t; \mathbf{x}))$, associated with the dynamic movement of $\mathbf{U}_0(t; \mathbf{x})$. This makes it possible to avoid any difficulty caused by diffusion in the movement of the autonomous object along the dynamic trajectory; this diffusion occurs because of a tangential component in the fluctuation excitation.

We shall restrict ourselves to a discussion of the simplest case when among the trajectories $\mathbf{U}_0(\mathbf{x})$ in the considered region $G^+(\mathbf{x})$ of phase space $G^+(\mathbf{x}) \subseteq G(\mathbf{x})$ there is a unique asymptotically stable ω-limit trajectory $U_0^+(x)$ [5] whose distance from any fairly close trajectory in Eq. (3) approaches zero as the phase increases. One of the movements $U_0^+(x)$ corresponding to $U_0^+(t+\tau_x; x)$ for suitable choice of τ_x is the dynamic approximation of the regulatable steady-state signal of the object $U(t; x)$. If the trajectory $U_0^+(x)$ does not degenerate into a point, then in the neighborhood $G(U_0^+(x))$, $G(U_0^+(x)) \subseteq G^+(x)$, by shifting from the Cartesian coordinates $U, \frac{dU}{dt}, \ldots, \frac{d^{q-1}U}{dt^{q-1}}$ to local coordinates associated with the motion of $U_0^+(t; x)$ (to the tangential phase-coordinate $t'(U)$ and the coordinates $U_0^+(x)$ which are orthogonal to the trajectory $n_k = n_k(U)$ (k = 1, 2, ..., q − 1)) we can use dynamic Eq. (3) to write an equivalent system of differential equations

$$\frac{dt}{dt'} + A_0(t', n_1, n_2, \ldots, n_{q-1}; x) = 0; \tag{5}$$

$$\frac{dn_m}{dt'} + A_m(t', n_1, n_2, \ldots, n_{q-1}; x) = 0 \quad (m = 1, 1, \ldots, q-1). \tag{6}$$

Assume that the function a in formula (4) is fairly smooth so that in a small neighborhood $G_1(U_0^+(x))$, $G_1(U_0^+(x)) \subseteq G(U_0^+(x))$, of the trajectory $U_0^+(x)$ only terms obtained by expanding the functions A_0 and A_m in powers of n_k (k = 1, 2, ..., q − 1) which are linear with respect to orthogonal deflections are of any great importance; here we obtain

$$\frac{d\sigma}{dt'} + \sum_{k=1}^{q-1} A_0^{(k)}(t'; x) n_k = 0; \tag{7}$$

$$\frac{dn_m}{dt'} + \sum_{k=1}^{q-1} A_m^{(k)}(t'; x) n_k = 0, \quad (m = 1, 2, \ldots, q-1), \tag{8}$$

where $\sigma \equiv t' - t(t'; x)$ is the phase shift. The fact that there are free terms in Eqs. (7) and (8) occurs because the motion of $U_0^+(t; x)$ is governed by the solutions of Eq. (3): the identities $t' \equiv t$, $n_k \equiv 0$ (k = 1, 2, ..., q − 1) must satisfy Eq. (3); this also means that they must satisfy Eqs. (7) and (8).

Three different sets of characteristic times are associated with the dynamic movements which are close to the nondegenerating trajectory $U_0^+(x)$. The first one $\{\Theta_{g_1}(x)\}$, is determined by dynamic motion along the trajectory $U_0^+(x)$ itself; for example, for the trajectory closed for all x when

$$U_0^+(t; x) \equiv U_0^+(t + T_x; x), \tag{9}$$

the set that $\theta_{g_1}(x)$ can contain values of the period T_x and several of its fractions corresponding to a limit motion rich in harmonics. The second set of times $\{\theta_{g_2}(x)\}$ determined by the values of $A_0^{(k)}(t'; x)$ and is an indicator of how isochronous the dynamic movements are near $U_0^+(x)$. The third set $\{\theta_g(x)\}$ describes the restoration of a dynamic system deflected in some manner in the neighborhood $G(U_0^+(x))$ from the limit trajectory $U_0^+(x)$ to this trajectory after the perturbation ceases. The last set $\{\theta_g(x)\}$ constitutes the set of relaxation times for the dynamic system near the ω-limit trajectory $U_0^+(x)$.

We set

$$\theta_m(x) \equiv \min_g \{\theta_g(x)\}, \qquad \theta_M(x) \equiv \max_g \{\theta_g(x)\}.$$

If $U_0^+(x)$ is a stationary point of Eq. (3), it is not necesary to shift from $U, \frac{dU}{dt}, \ldots, \frac{d^{q-1}U}{dt^{q-1}}$ to new coordinates; linearization of the dynamic equation is performed directly with $U - U^+$, $\frac{dU}{dt}, \ldots, \frac{d^{q-1}U}{dt^{q-1}}$. In this case all q characteristic times are relaxation times for the dynamic system to the state of asymptotically stable equilibrium.

According to the above, we shall begin with an approximation of the observed signal U(t; x) using the equation

$$D_0[U(t; x)] = F(t; x), \tag{10}$$

in which for each value of **x**, the term F(t; x) is a stationary random process with zero mathematical expectation: $\langle F(t; x) \rangle \equiv 0$. In all physical examples (when the source of signal fluctuations is: 1) thermal noise in a system in thermodynamic equilibrium; 2) shot effect in particle streams; 3) noise induced by spontaneous transition in a quantum mechanical system) analyzed from the viewpoint of the structure of forces perturbing the corresponding dynamic system, we can treat the function F(t; x) as δ-correlated Gauss fluctuations: $\langle F(t; x) F(t + \tau; x) \rangle \equiv C(t; x)\delta(\tau)$. It is natural to expect that a similar situation in which the change in U(t; x) with time is due to the interrelated motion of a large number of the same kind of microparts of the object occurs not only in natural-science problems but in, for example, economics as well. For this reason, without making any additional refinement or justifications, we assume that the values of the fluctuating perturbation completely lose their stochastic relationship for time shifts $|\tau| \geq \tau_0$, where the magnitude of the minimum value of τ_0, characterizing the duration of the stochastic relationship for the fluctuating excitation is considerably less than the minimum relaxation time $\theta_m(x)$ of dynamic system (3). Below, assuming F(t; x) is a Gaussian random process, we begin with the relationships

$$\langle F(t; x) F(t+\tau; x) \rangle = 0 \quad \text{for} \quad |t|, |t+\tau| \leqslant \frac{\theta}{2}, \quad |\tau| \gg \tau_0, \quad \tau_0 \ll \theta_m(x). \tag{11}$$

Despite the presence of a fluctuation excitation explicitly containing the time in Eq. (10) we shall continue to characterize the observed state of the object by the quantities $\mathbf{U} \equiv \left(U, \frac{dU}{dt}, \ldots, \frac{d^{q-1}U}{dt^{q-1}}\right)$ and describe the change in the signal with time by the positions of the image point U(t; x) in the phase space $R_q(x)$ of an unexcited dynamic system. Assuming that the observed changes in the signal are fairly well approximated by dynamic Eq. (3) we can assume that the response rate of the corresponding dynamic system to the fluctuation excitation F(t; x) is fairly low. For this reason we shall assume that internal fluctuations do not remove the steady-state signal from the neighborhood of the asymptotically stable ω-limit trajectory of the dynamic system. For a nondegenerating trajectory $\mathbf{U}_0^+(x)$ we have

$$\frac{dt}{dt'} + A_0(t', n_1, n_2, \ldots, n_{q-1}; x) = F_0(t'; x); \tag{12}$$

$$\frac{dn_m}{dt'} + A_m(t', n_1, n_2, \ldots, n_{q-1}; x) = F_m(t'; x) \quad (m = 1, 2, \ldots, q-1), \tag{13}$$

where $F_0(t'; x)$ and $\{F_m(t'; x)\}$ are the components of the fluctuation excitation (scaled to the new argument t') along the trajectory $\mathbf{U}_0^+(x)$ and in different (m = 1, 2, ..., q − 1) directions orthogonal to $\mathbf{U}_0^+(x)$:

$$\langle F_m(t'; x) \rangle \equiv 0, \, (m = 0, 1, \ldots, q-1).$$

The stochastic relationship between the values of the components, time-shifted by τ disappears for shifts $|\tau| \geq \tau$.

If the fluctuations are so small that the image points of the object during the signal observation interval $(-\Theta/2, \Theta/2)$ lie almost completely within the neighborhood $G_1(U_0(\mathbf{x}))$, the equations for the signal take the form

$$\frac{d\sigma}{dt'} + \sum_{k=1}^{q-1} A_0^{(k)}(t'; \mathbf{x}) n_k = F_0(t'; \mathbf{x}); \tag{14}$$

$$\frac{dn_m}{dt'} + \sum_{l=1}^{q-1} A_m^{(k)}(t'; \mathbf{x}) n_k = F_m(t'; \mathbf{x}) \quad (m = 1, 2, \ldots, q-1). \tag{15}$$

In this case, after calculating the position of the ω-limit dynamic trajectory $U_0^+(\mathbf{x})$ in phase space $R_q(\mathbf{x})$, the problem of analyzing the dynamic structure of the object for each value of $\mathbf{x}$ reduces to estimating the numerical values of the coefficients A_0^k, $A_m^{(k)}$ (k, m = 1, 2, ..., q − 1).

Even the presence of a limit trajectory not degenerating into a stationary point indicates that dynamic system (3) is nonlinear. Therefore, linearization of Eqs. (5)-(8) near the non-degenerating trajectory $U_0^+(\mathbf{x})$ does not reduce the dynamic equation to a linear one, but does simplify and restrict it. However, if the internal fluctuations are so small that nonlinear [with respect to deviations from the trajectory $U_0^+(\mathbf{x})$] effects are not reliably detected, the signal will not carry more detailed information about the operator $D_0[U; \mathbf{x}]$ in the general case. Only when *a priori* information allows us to isolate a sufficiently narrow class of operators $\{D_\alpha\}$ does the recording of a signal not leaving the neighborhood $(-\Theta/2, \Theta/2)$ during a time interval $G_1(U_0^+(\mathbf{x}))$ make it possible to reveal the dynamic structure of an object with all the necessary details.*

In accordance with formulas (3), (4), and (10), for motion due to the effect of a fluctuation perturbation in the neighborhood of an isolated asymptotically stable stationary point $U = U_0^+(\mathbf{x})$ of dynamic system (3), we can write

$$\frac{d^q U}{dt^q} + a\left(U, \frac{dU}{dt}, \ldots, \frac{d^{q-1}U}{dt^{q-1}}; \mathbf{x}\right) = F(t; \mathbf{x}). \tag{16}$$

If the internal fluctuations due not remove the image point of the object from the neighborhood $G_1(U_0^+(\mathbf{x}))$, where only the linear terms obtained by expanding the deviations $U - U_0^+$, $\frac{dU}{dt}$, $\ldots, \frac{d^{q-1}U}{dt^{q-1}}$ from the stationary point in a power series are important in the function a, the equation of the signal takes the form

$$\frac{d^q U}{dt^q} + a_0(\mathbf{x})[U - U_0^+(\mathbf{x})] + \sum_{k=1}^{q-1} a_k(\mathbf{x}) \frac{d^k U}{dt^k} = F(t; \mathbf{x}). \tag{17}$$

§3. Method of Obtaining the Dynamic Operator

We shall consider one possible method for estimating the coefficients $A_m^{(k)}(\mathbf{x})$ or $a_k(\mathbf{x})$ for the dynamic system of an object linearized with respect to the deviations from an isolated

---

*Obviously, expressions of the type "recording of object structure" cannot be taken literally since signals recorded in practice always contain only limited information about a particular characteristic of the object.

limit trajectory $\mathbf{U}_0^+(\mathbf{x})$ which nowhere compactly fills the two-dimensional x surface in region $G^+(\mathbf{x})$. First consider the variant in which $\mathbf{U}_0^+(\mathbf{x})$ does not degenerate into a stationary point. Let the description of the signal $U(t;\mathbf{x})$ on the interval $-\Theta/2 \leq t < \Theta/2$ contain a large number Γ, $(\Gamma \gg 1)$ of stochastically independent realizations $U_\gamma(t;\mathbf{x})$ $(t_{\gamma 1} \leq t < t_{\gamma 2}, \gamma = 1, 2, \ldots, \Gamma)$. For simplicity, assume that each of these realizations, for fixed x, passes through the same segment $t'_1(\mathbf{x}) \leq t' < t'_2(\mathbf{x})$ of phase change, read off along the trajectory $\mathbf{U}_0^+(\mathbf{x})$ from any point $\mathbf{U}_0(\mathbf{x})$. In this case, for a low internal-fluctuation intensity, the observed motion of $U_\gamma(t;\mathbf{x})$ repeatedly falls in a fairly narrow neighborhood $G^*(\mathbf{U}_0^+(\mathbf{x}))$, $G^*(\mathbf{U}_0^+(\mathbf{x})) \subseteq G(\mathbf{U}_0^+(\mathbf{x}))$, and the segment (t'_1, t'_2) of the trajectory $\mathbf{U}_0^+(\mathbf{x})$ remains almost completely within the limit G^* for $t'_1 \leq t' < t'_2$. This makes it possible to estimate the position of the mean statistical trajectory $\mathbf{U}^*(\mathbf{x})$ in space $R_q(\mathbf{x})$ on the interval (t'_1, t'_2) for which the mathematical expectation of the deviations in the ensemble of trajectories of the fluctuating object which are close to $\mathbf{U}_0^+(\mathbf{x})$ becomes zero. Calculation will be performed starting from the trajectory of any realization $\mathbf{U}_{\gamma_1}(\mathbf{x})$ i.e., from the first approximation: $\mathbf{U}^{(1)}(\mathbf{x}) \equiv \mathbf{U}_{\gamma_1}(\mathbf{x})$. The method of successive approximations is used as follows: orthogonal hyperplanes are drawn through different points on the trajectory $\mathbf{U}_{(k)}(\mathbf{x})$ of the k-th approximation. The trajectory passing through the centroid of the points where these hyperplanes intersect each of the trajectories $\mathbf{U}_\gamma(\mathbf{x})$, is taken as the $(k+1)$-th approximation of the mean statistical trajectory $\mathbf{U}^{(k+1)}(\mathbf{x})$. The limit trajectory $\mathbf{U}^{(k)}(\mathbf{x})$ for the sequence $\mathbf{U}^{(m)}(\mathbf{x})$ is characterized by the fact that the mean value of the orthogonal deviations from its realizations $\mathbf{U}_\gamma(\mathbf{x})$ becomes zero; thus, we can assume that $\mathbf{U}^{(\infty)}(\mathbf{x})$ is an estimate of the trajectory $\mathbf{U}^*(\mathbf{x})$.

When the rate of internal fluctuations in the object is so low that the observed realizations $G_1(U_0(\mathbf{x}))$ do not lie within the neighborhood $G_1(\mathbf{U}_0^+(\mathbf{x}))$, where the behavior of the signal is described by Eqs. (15) linearized in $(n_1, n_2, \ldots, n_{q-1})$, the ω-limit trajectory of dynamic system $\mathbf{U}_0^+(\mathbf{x})$ (for which $n_k(t) \equiv 0$, $k = 1, 2, \ldots, q-1$) coincides with the mean statistical trajectory; the orthogonal deviations $n_k^*(t')$ in the points of this trajectory, read off from $\mathbf{U}_0^+(\mathbf{x})$ are determined by the formulas $n_k^*(t') \equiv \langle n(t') \rangle$. Now by knowing the position of the trajectory $\mathbf{U}_0^+(\mathbf{x})$ in the phase space $R_q(\mathbf{x})$ where the Cartesian coordinates are $U, \frac{dU}{dt}, \ldots, \frac{d^{q-1}U}{dt^{q-1}}$, we can mark the time for dynamic motion along $\mathbf{U}_0^+(\mathbf{x})$ by knowing the state of U_0^+ in addition to the initial one. Then, in the neighborhood $G(\mathbf{U}_0^+(\mathbf{x}))$, we find the functions $U_\gamma(t;\mathbf{x})$ and $n_{k\gamma}(t';\mathbf{x})$, $(k = 1, 2, \ldots, q-1)$ for each realization $\sigma_\gamma(t';\mathbf{x})$, in accordance with the above assumption these functions are stochastically independent for different γ.

We consider a calculation method making it possible to estimate the values of the coefficients $A_0^{(k)}$ in Eq. (14). We rewrite this equation for any realization $U_\gamma(t;\mathbf{x})$ in the more convenient form

$$\frac{d\sigma_\gamma}{dt'}(t'+\tau;\mathbf{x}) + \sum_{k=1}^{q-1} A_0^{(k)}(t'+\tau;\mathbf{x}) n_{k\gamma}(t'+\tau;\mathbf{x}) = F_{0\gamma}(t'+\tau;\mathbf{x}) \quad (\gamma = 1, 2, \ldots, \Gamma). \tag{14'}$$

By multiplying (14') by $n_{l\gamma}(t',\mathbf{x})$ $(l = 1, 2, \ldots, q-1)$ and introducing the notation

$$\chi_{\Gamma 0 l}^{(t';\mathbf{x})}(\tau) \equiv \frac{1}{\Gamma} \sum_{\gamma=1}^{\Gamma} n_{l\gamma}(t';\mathbf{x}) \frac{d\sigma_\gamma}{dt'}(t'+\tau;\mathbf{x}), \tag{14''}$$

$$\eta_{\Gamma k l}^{(t';\mathbf{x})}(\tau) \equiv \frac{1}{\Gamma} \sum_{\gamma=1}^{\Gamma} n_{l\gamma}(t';\mathbf{x}) n_{k\gamma}(t'\,\mathbf{x}), \quad \varphi_{\Gamma 0 l}^{(t;\mathbf{x})}(\tau) \equiv \frac{1}{\Gamma} \sum_{\gamma=1}^{\Gamma} n_{l\gamma}(t';\mathbf{x}) F_{0\gamma}(t';\mathbf{x}),$$

we obtain the formulas

$$\chi_{\Gamma 0l}^{(t';\mathbf{x})}(\tau) + \sum_{k=1}^{q-1} A_0^{(k)}(t'+\tau;\mathbf{x})\eta_{\Gamma kl}^{(t';\mathbf{x})}(\tau) = \varphi_{\Gamma 0l}^{(t';\mathbf{x})}(\tau) \quad (l=1,2,\ldots,q-1), \tag{18}$$

which associate the sample correlation functions. As already noted, the stochastic relationship between the time-shifted values of the fluctuation excitation F which apply to the same value of **x** disappears for shifts greater than τ_0. Since the response of the dynamic system to a perturbing force is determined only by the preceding values of the force, the stochastic relationship between the signal U or its projections σ and n_l ($l = 1, 2, \ldots, q-1$) and the force F or its projections F_m ($m = 0, 1, \ldots, q-1$) disappears when the response precedes the force by a time greater than τ_0. In particular,

$$\langle n_l(t';\mathbf{x}) F_m(t';\mathbf{x}) \rangle = 0,$$
$$\tau \gg \tau_0 \, (l=1,2,\ldots,q-1; \quad m=0,1,\ldots,q-1).$$

It now follows that

$$\chi_{0l}^{(t';\mathbf{x})}(\tau) + \sum_{k=1}^{q-1} A_0^{(k)}(t'+\tau;\mathbf{x})\eta_{kl}^{(t';\mathbf{x})}(\tau) = 0, \quad \tau \gg \tau_0 \, (l=1,2,\ldots,q-1), \tag{20}$$

where

$$\chi_{0l}^{(t';\mathbf{x})}(\tau) \equiv \langle n_l(t'+\tau;\mathbf{x}) \frac{d\sigma}{dt'}(t';\mathbf{x}) \rangle,$$

$$\eta_{lk}^{(t';\mathbf{x})}(\tau) \equiv \langle n_l(t'+\tau;\mathbf{x}) n_k(t';\mathbf{x}) \rangle.$$

In addition, since

$$\langle [\varphi_{\Gamma 0l}^{(t';\mathbf{x})}(\tau)]^2 \rangle = \frac{1}{\Gamma} \langle [n_l(t';\mathbf{x}) F_0(t'+\tau;\mathbf{x})]^2 \rangle, \tag{21}$$

we can estimate the coefficients $A_0^{(k)}(t';\mathbf{x})$ in terms of the quantities $A_0^k(t',\mathbf{x};\tau)$ to a zeroth approximation using the approximate formulas

$$\chi_{\Gamma 0l}^{(t'-\tau;\mathbf{x})}(\tau) + \sum_{k=1}^{q-1} \mathcal{A}_0^{(k)}(t';\mathbf{x};\tau)\eta_{\Gamma kl}^{(t'-\tau;\mathbf{x})}(\tau) = 0, \tag{22}$$

where τ is the shift within the interval $\tau_0 \le \tau < \theta_m(\mathbf{x})$. In fact, the random quantities $\varphi_{\Gamma 0l}^{(t';\mathbf{x})}{}_{(\tau)}$ which are dropped from the right side of Eqs. (18) when writing (22) have zero mathematical expectations, according to (20) and (21), and approach zero as the dispersion $\tau \ge \tau_0$ increases. The result of this estimate is the random function τ.

In order to make complete use of the information contained in the recorded signal it is natural to estimate the coefficients $A_0^{(k)}$ by minimizing the rms error due to the quantities $\varphi_{\Gamma 0l}^{(t';\mathbf{x})}{}_{(\tau)}$. However, it is difficult to directly apply the technique usually used in such estimates to Eqs. (18) for $t'+\tau = \theta$ and different τ in order to obtain relationships associating [like (22)] the coefficients for identical phases (θ) because of the strong stochastic relationship between the values $\varphi_{\Gamma 0 l}^{(t';\mathbf{x})}{}_{(\tau)}$. In fact,

$$\langle \varphi_{\Gamma 0 l}^{(\theta-\tau_1;\mathbf{x})}(\tau_1) \varphi_{\Gamma 0 k}^{(\theta-\tau_2;\mathbf{x})}(\tau_2) \rangle = \frac{1}{\Gamma} \langle n_l(\theta-\tau_1;\mathbf{x}) n_k(\theta-\tau_2;\mathbf{x}) [F_0(\theta;\mathbf{x})]^2 \rangle.$$

For large Γ the distribution of the vector random process $\varphi_{\Gamma 0 l}^{(\theta-\tau;\mathbf{x})}(\tau)$ ($l=1, 2, \ldots, q-1$) approaches a Gaussian law; by allowing for this when $\tau_1, \tau_2 \geq \tau_0$, we can rewrite the last expression for its correlation matrix as

$$\langle \varphi_{\Gamma 0 l}^{(\theta-\tau_1;\mathbf{x})}(\tau_1) \varphi_{\Gamma 0 k}^{(\theta-\tau_2;\mathbf{x})}(\tau_2) \rangle = \frac{1}{\Gamma} \langle [F_0(\theta;\mathbf{x})]^2 \rangle \langle n_l(\theta-\tau_1;\mathbf{x}) n_k(\theta-\tau_2;\mathbf{x}) \rangle. \tag{23}$$

Thus, when all components of matrix (23) approach zero in almost the same way and there is a corresponding disappearance of a stochastic relationship between Γ for large $\varphi_{\Gamma 0 l}^{(\theta-\tau;\mathbf{x})}{}_{(\tau)}$ it is necessary to have large shifts so that the matrix $\eta_{k l}^{(\theta-\tau_1)}{}_{(\tau_2-\tau_1;)}\mathbf{x}$ disappears as well. However, for these shifts, it is clear from Eqs. (18) that it is unreasonable to estimate the coefficients A_0^k; to smaller values of $|\tau_1 - \tau_2|$ there correspond components of the correlation matrix (23) which decay as $1/\Gamma$, i.e., according to (21) they decay at the same rate as the dispersion $\varphi_{\Gamma 0 l}^{(\theta-\tau;\mathbf{x})}{}_{(\tau)}$.

To aid in this situation we assume that our *a priori* representations for the dynamic structure of the analyzable object make it possible to represent the sought coefficients in the form of single-values expansions

$$A_0^{(k)}(t';\mathbf{x}) \equiv \sum_{r=1}^{R} A_0^{(k,r;\mathbf{x})} v_{0r}(t';\mathbf{x}) \tag{24}$$

in powers of previously known functions. Note that in principle we can now check assumption (24) because the estimate of the quantities $A_0^k(t';\mathbf{x})$ obtained in the zeroth approximation from formulas (22) is not needed in *a priori* information of this type. By substituting Eq. (24) into Eq. (18) and letting $\tau = \tau_p$, $\tau_p = p\tau_0$ ($p=1, 2, \ldots, P$; $P \approx \theta_m/\tau_0$) in the latter we obtain the following system of equations for estimating the coefficients $A_0^{(k,r;\mathbf{x})}$:

$$\chi_{\Gamma 0 l}^{(t';\mathbf{x})}(\tau_p) + \sum_{k=1}^{q-1} \sum_{r=1}^{R} A_0^{(k,r;\mathbf{x})} v_{0r}(t'+\tau_p;\mathbf{x}) \eta_{\Gamma k l}^{(t';\mathbf{x})}(\tau_p) = \varphi_{\Gamma 0 l}^{(t';\mathbf{x})}(\tau_p), \tag{25}$$

where we should take $t' = t'_1, t'_1 + P\tau_0, t'_1 + 2P\tau_0, \ldots, t'_1 + \varkappa P\tau_0$;

$$t'_1 + \varkappa P\tau_0 \simeq t'_2; \quad p = 1, 2, \ldots, P; \quad kl = 1, 2, \ldots, q-1;$$
$$r = 1, 2, \ldots, R.$$

We assume the functions $v_{0r}(t^0;\mathbf{x})$ are known and the values of $\chi_{\Gamma 0 l}^{(t'\mathbf{x})}(t_p)$ and $\eta_{\Gamma k l}^{(t'\mathbf{x})}(\tau_p)$ are found, according to (14) from the description of the signal. The right sides of these equations now contain quantities which are stochastically independent for different p. In fact, for sufficiently large Γ we can assume the fluctuations $\varphi_{\Gamma 0 l}^{(t';\mathbf{x})}(\tau)$ are distributed according to a normal law and, as is not difficult to verify by direct substitution, there is no correlation between them for different p (p = 1, 2, ..., P):

$$\langle \varphi_{\Gamma 0 l}^{(t';\mathbf{x})}(\tau') \varphi_{\Gamma 0 k}^{(t';\mathbf{x})}(\tau'') \rangle = 0 \qquad |\tau' - \tau''| \geqslant \tau_0. \tag{26}$$

Therefore, by using system (25) as a starting point, we can estimate the coefficients $A_0^{(k,r;\mathbf{x})}$, by using the current technique for finding the minimum in the rms error [6]. In particular, the coefficients $A_0^{(k,r,\mathbf{x})}$ are estimated by the quantities $\hat{A}_0^{(k,r,\mathbf{x})}$, realizing a minimum for the functional

$$\Phi(\mathcal{A}_0^{(k,r;\mathbf{x})}; \ k = 1, 2, \ldots, q-1; \ r = 1, 2, \ldots, R) \equiv$$
$$\equiv \sum_{t',p,l} \frac{\left[\chi_{\Gamma 0 l}^{(t';\mathbf{x})}(\tau_p) + \sum_{k=1}^{q-1} \sum_{r=1}^{R} \mathcal{A}_0^{(k,r;\mathbf{x})} v_{0r}(t'+\tau_p;\mathbf{x}) \eta_{\Gamma k l}^{(t';\mathbf{x})}(\tau_p) \right]^2}{g_{0 t' p l}^{(\mathbf{x})}(\mathcal{A}_0^{(k,r;\mathbf{x})}; \ k=1, 2, \ldots, q-1; \ r=1, 2, \ldots, R)}. \tag{27}$$

The expressions

$$g_{0 t' p l}^{(\mathbf{x})} \equiv \frac{1}{\Gamma} \sum_{\gamma=1}^{\Gamma} \left[\chi_{0 l \gamma}^{(t',\mathbf{x})}(\tau_p) + \sum_{k=1}^{q-1} \sum_{r=1}^{R} \mathcal{A}_0^{(k,r;\mathbf{x})} v_{0r}(t'+\tau_p;\mathbf{x}) \eta_{k l \gamma}^{(t';\mathbf{x})}(\tau_p) \right]^2, \tag{28}$$

appearing in (27) where

$$\chi_{0 l \gamma}^{(t';\mathbf{x})}(\tau) \equiv n_{l\gamma}(t';\mathbf{x}) \frac{d\sigma_\gamma}{dt'}(t'-\tau;\mathbf{x}), \qquad \eta_{k l \gamma}^{(t';\mathbf{x})}(\tau) \equiv n_{l\gamma}(t';\mathbf{x}) n_{k\gamma}(t'-\tau;\mathbf{x})$$

estimate the dispersion in the "fluctuation errors" $\varphi_{\Gamma 0 l}^{(t';\mathbf{x})}(\tau)$. It is extremely difficult to find the values $\hat{\mathcal{A}}_0^{(k,r;\mathbf{x})}$ by setting the partial derivatives with respect to $\mathcal{A}_0^{(k,r;\mathbf{x})}$ of the function $\Phi_0^{(\mathbf{x})}$ equal to zero allowing for (28).

Its solution can also be obtained by successive linear approximations; these can be formulated as follows: let the quantities $\hat{\mathcal{A}}_{0\mu}^{(k,r,\mathbf{x})}$ realize a minimum in the μ-th ($\mu = 1, 2, \ldots$) functional $\Phi_{0\mu}^{(\mathbf{x})}$

$$\Phi_{0\mu}^{(\mathbf{x})} \equiv \sum_{t',p,l} \frac{1}{g_{0\mu-1}^{(\mathbf{x})}(t',p,l)} \left[\chi_{\Gamma 0 l}^{(t';\mathbf{x})}(\tau_p) + \sum_{k=1}^{q-1} \sum_{r=1}^{R} \mathcal{A}_{0\mu}^{(k,r;\mathbf{x})} v_{0r}(t'+\tau_p;\mathbf{x}) \eta_{\Gamma k l}^{(t';\mathbf{x})}(\tau_p) \right]^2, \tag{29}$$

where the values of the dispersion in the errors $g_{0\mu-1}^{(\mathbf{x})}$ are estimated using the solution of the preceding approximation:

$$g_{0\mu-1}^{(\mathbf{x})}(t',p,l) \equiv \frac{1}{\Gamma} \sum_{\gamma=1}^{\Gamma} \left[\chi_{0 l \gamma}^{(t';\mathbf{x})}(\tau_p) + \sum_{k=1}^{q-1} \sum_{r=1}^{R} \hat{\mathcal{A}}_{0\mu-1}^{(k,r;\mathbf{x})} v_{0r}(t'+\tau_p;\mathbf{x}) \eta_{k l \gamma}^{(t';\mathbf{x})}(\tau_p) \right]^2. \tag{30}$$

For $\mu = 1$ we should substitute the values of $A_{00}^{k,r;\mathbf{x}}$ obtained from (22) and (24) into expressions (29) and (30); finally, we can simply let $g_{00}^{(\mathbf{x})}(t', p, l) \equiv 1$.

The components $A_m^{(k)}$ of the matrix of coefficients for system (15) are estimated similarly. Here, instead of the random vector $\chi_{\Gamma 0l}^{(t',\mathbf{x})}(\tau)$ $(l=1, 2, \ldots, q-1)$ we use the random matrix $\|\chi_{\Gamma ml}^{(t';\mathbf{x})}(\tau)\|$, where

$$\chi_{\Gamma ml}^{(t';\mathbf{x})}(\tau) \equiv \frac{1}{\Gamma} \sum_{\gamma=1}^{\Gamma} n_{l\gamma}(t', \mathbf{x}) \frac{dn_{m\gamma}}{dt'}(t' + \tau; \mathbf{x}) \quad (l, m = 1, 2, \ldots, q-1), \tag{31}$$

and in place of the fluctuation-error vector $\varphi_{\Gamma 0l}^{(t';\mathbf{x})}(\tau)$ $(l = 1, 2, \ldots, q-1)$ we use the matrix $\|\varphi_{\Gamma ml}^{(t';\mathbf{x})}(\tau)\|$, in which

$$\varphi_{\Gamma ml}^{(t';\mathbf{x})}(\tau) \equiv \frac{1}{\Gamma} \sum_{\gamma=1}^{\Gamma} n_{l\gamma}(t';\mathbf{x}) F_{m\gamma}(t' + \tau;\mathbf{x}). \tag{32}$$

Thus, from a fairly long and detailed description of the signal U(t; **x**) we can estimate the dynamic characteristics of the weakly fluctuating nonregulatable object (very general type) near its ω-limit trajectory (not degenerating into a stationary point) by performing some complicated numerical operations on an electronic computer.

We shall consider in somewhat more detail a simple and important example of a non-degenerating $\overline{U_0^+(\mathbf{x})}$. Let the ω-limit trajectory of the dynamic system of the studied object represent a closed curve for all values of the paramters $x_1 \in X_i$ $(i = 1, 2, \ldots, Q)$. Then to each value of **x** there corresponds a positive number $T_\mathbf{x}$ — the period of dynamic motion — for which, in particular,

$$U_0^+(t + T_\mathbf{x}; \mathbf{x}) \equiv U_0^+(t; x) \tag{33}$$

All of the remaining dynamic characteristics $A_m^{(k)}(t'; \mathbf{x})$ of the dynamic system linearized to the deviations from the limit cycle will clearly also be periodic with period $T_\mathbf{x}$:

$$A_m^{(k)}(t' + T_\mathbf{x}; \mathbf{x}) \equiv A_m^{(k)}(t'; \mathbf{x}), \quad (m = 0, 1, \ldots, q-1; \quad k = 1, 2, \ldots, q-1). \tag{34}$$

The system of functions $v_{mr}(t'; \mathbf{x})$ which are used in the expansion of these characteristics consists of the following harmonics:

$$v_{mr}(t'; \mathbf{x}) \equiv \exp\left(i \frac{2\pi r}{T_\mathbf{x}} t'\right)(r = 0, \pm 1, \ldots, \pm R; m = 0, 1, \ldots, q-1).$$

From what has been said, the equations of the successive ($\mu = 1, 2, \ldots$) approximations (29), (30) for the estimates $\hat{A}_{m\mu}^{(k,r;\mathbf{x})}$ of the Fourier coefficients $A_m^{k,r;\mathbf{x}}$ of the functions

$$A_m^{(k)}(t'; \mathbf{x}) \equiv \sum_{r=-R}^{R} A_m^{(k,r;\mathbf{x})} \exp\left(i \frac{2\pi r}{T_\mathbf{x}} t'\right)$$

can be represented in a more specific form:

$$\sum_{p=1}^{P} \exp\left(i\frac{2\mu p r \tau_0}{T_\mathbf{x}}\right) \sum_{l=1}^{q-1} \left\{ \int_0^{T_\mathbf{x}} \exp\left(i\frac{2\pi r t'}{T_\mathbf{x}}\right) [g_{m,\mu-1}^{(\mathbf{x})}(t', p, l)]^{-1} \eta_{\Gamma k l}^{(t';\,\mathbf{x})}(p\tau_0) \chi_{\Gamma m l}^{(t';\,\mathbf{x})}(p\tau_0) dt' + \right.$$

$$\left. + \sum_{\sigma=-R}^{R} \exp\left(i\frac{2\pi p \sigma \tau_0}{T_\mathbf{x}}\right) \sum_{\beta=1}^{q-1} \hat{\mathcal{A}}_{m\mu}^{(\beta,\,\sigma;\,\mathbf{x})} \int_0^{T_\mathbf{x}} \exp\left[i 2\pi \frac{r+\sigma}{T_\mathbf{x}} t'\right] [g_{m\mu-1}^{(\mathbf{x})}(t', p, l)]^{-1} \eta_{\Gamma k l}^{(t';\,\mathbf{x})}(p\tau_0) \eta_{\Gamma \beta l}^{(t';\,\mathbf{x})}(p\tau_0) dt' \right\} = 0,$$

$$g_{m\mu-1}^{(\mathbf{x})}(t', p, l) \equiv \frac{1}{\Gamma} \sum_{\gamma=1}^{\Gamma} \left\{ \chi_{m l \gamma}^{(t';\,\mathbf{x})}(p\tau_0) + \sum_{\beta=1}^{q-1} \eta_{\beta l \gamma}^{(t';\,\mathbf{x})}(p\tau_0) \sum_{\sigma=-R}^{R} \hat{\mathcal{A}}_{m\mu-1}^{(\beta,\,\sigma;\,\mathbf{x})} \exp\left[i\frac{2\pi\sigma}{T_\mathbf{x}}(t^2 + p\tau_0)\right] \right\}^2$$

$$(m = 0, 1, \ldots, q-1; \quad l, k = 1, 2, \ldots, q-1; \quad p = 1, 2, \ldots, P;$$
$$r = 0 \pm 1, \pm 2, \ldots, \pm R). \tag{35}$$

The values of the estimates $\hat{\mathcal{A}}_m^{(k,p;x)}$ in the zeroth approximation can be found from the equation

$$\int_0^{T_\mathbf{x}} \chi_{\Gamma m l}^{(t';\,\mathbf{x})}(\tau_0) \exp\left(i\frac{2\pi\sigma t'}{T_\mathbf{x}}\right) dt' + \sum_{\beta=1}^{q-1} \sum_{\rho=-R}^{R} \hat{\mathcal{A}}_{m0}^{(\beta,\,\rho;\,\mathbf{x})} \exp\left(i\frac{2\pi\rho\tau_0}{T_\mathbf{x}}\right) \int_0^{T_\mathbf{x}} \eta_{\Gamma \beta l}^{(t';\,\mathbf{x})}(\tau_0) \exp\left[i\frac{2\pi(\sigma+\rho)t'}{T_\mathbf{x}}\right] dt' = 0$$

$$(m = 0, 1, \ldots, q-1; \quad l = 1, 2, \ldots, q-1; \quad \sigma = 0 \pm 1, \pm 2, \ldots, \pm R).$$

We now discuss the simplest variant of the problem of analyzing a steady-state signal. Let the studied object move under the effect of internal fluctuations near an asymptotically stable stationary point $U_0^+(\mathbf{x})$ of its dynamic system while still remaining within the neighborhood $G_1(U_0^+(\mathbf{x}))$ for the entire observation time. Without introducing any other notation, we can write the observed signal as the linear equation

$$\frac{d^q U}{dt^q}(t;\mathbf{x}) + \sum_{k=0}^{q-1} a_k(\mathbf{x}) \frac{d^k U}{dt^k}(t;\mathbf{x}) = F(t;\mathbf{x}), \tag{17'}$$

whose constants $a_k(\mathbf{x})$ (k = 0, 1, ..., q − 1) must be estimated. To do this we reason as before, although in this case the problem is simpler. We divide the observation $[-\Theta/2 \leq t < \Theta/2]$ into segments of length $\Theta_0 = N\tau_0$ for each of which $N \gg q$, $\Theta_0 \lessgtr \theta_m$. Here we assume that $[\Gamma \equiv \Theta/\Theta_0 \gg 1]$.

We set

$$U_\gamma^{(\mathbf{x})}(t) \equiv U(t_\gamma + t; \mathbf{x}), \qquad F_\gamma^{(\mathbf{x})}(t) \equiv F(t_\gamma + t; \mathbf{x}), \qquad 0 \leq t < \Theta_0,$$

$$t_\gamma = -\frac{\Theta}{2} + (\gamma - 1)\Theta_0;$$

$$\chi_{(\gamma)}^{(t;\,\mathbf{x})}(\tau) \equiv U_\gamma^{(\mathbf{x})}(t) U_\gamma^{(\mathbf{x})}(t+\tau), \qquad \varphi_{(\gamma)}^{(t;\,\mathbf{x})}(\tau) \equiv U_\gamma^{(\mathbf{x})}(t) F_\gamma^{(\mathbf{x})}(t+\tau)$$

$$(\gamma = 1, 2, \ldots, \Gamma).$$

In view of $[U_\gamma^{(x)}(t)\frac{d^k}{dt^k}[U_\gamma^{(x)}(t+\tau)] \equiv \frac{\partial^k}{\partial \tau^k}\chi_{(\gamma)}^{(t;\,x)}(\tau)]$ Eq. (17') yields

$$\left[\frac{\partial^q}{\partial \tau^q} + \sum_{k=0}^{q-1} a_k(\mathbf{x})\frac{\partial^k}{\partial \tau^k}\right]\chi_{(\gamma)}^{(t;\,\mathbf{x})}(\tau) = \varphi_{(\gamma)}^{(t;\,\mathbf{x})}(\tau). \tag{36}$$

We introduce the sample correlation functions

$$\chi_\Gamma^{(t;\,\mathbf{x})} \equiv \frac{1}{\Gamma}\sum_{\gamma=1}^{\Gamma}\chi_{(\gamma)}^{(t;\,\mathbf{x})}(\tau), \qquad \varphi_\Gamma^{(t;\,\mathbf{x})} \equiv \frac{1}{\Gamma}\sum_{\gamma=1}^{\Gamma}\varphi_{(\gamma)}^{(t;\,\mathbf{x})}(\tau).$$

Omitting the subscripts and setting $\tau = \sigma \tau_0$, $t = 0$, we obtain

$$\left[\frac{d^q}{d\tau^q} + \sum_{k=0}^{q-1} a_k(\mathbf{x})\frac{d^k}{d\tau^k}\right]\chi(\sigma\tau_0) = \varphi(\sigma\tau_0) \quad (\sigma = 1, 2, \ldots, N). \tag{37}$$

As before, we have $\langle \varphi(\sigma\tau_0)\rangle = 0$ in view of the considerations used in deriving (19). Allowing for the normality of the random processes $F_\gamma(t)$ and $U_\gamma(t)$, we have

$$\langle \varphi(\sigma'\tau_0)\varphi(\sigma''\tau_0)\rangle = \frac{1}{\Gamma^2}\sum_{\gamma'=1}^{\Gamma}\sum_{\gamma''=1}^{\Gamma}\langle U_{\gamma'}(t)U_{\gamma''}(t)F_{\gamma'}(t+\sigma'\tau_0)F_{\gamma''}(t+\sigma'\tau_0)\rangle =$$

$$= \frac{1}{\Gamma^2}\sum_{\gamma'=1}^{\Gamma}\sum_{\gamma''=1}^{\Gamma}[\langle U_{\gamma'}(t)F_{\gamma'}(t+\sigma'\tau_0)\rangle\langle U_{\gamma''}(t)F_{\gamma''}(t+\sigma''\tau_0)\rangle +$$

$$+ \langle U_{\gamma'}(t)F_{\gamma''}(t+\sigma''\tau_0)\rangle\langle U_{\gamma''}(t)F_{\gamma'}(t+\sigma'\tau_0)\rangle +$$

$$+ \langle U_{\gamma'}(t)U_{\gamma''}(t)\rangle F_{\gamma'}(t+\sigma'\tau_0)F_{\gamma''}(t+\sigma''\tau_0)\rangle] = 0.$$

In correspondance with the asymptotic normality of the process $\Gamma \gg 1$, when $\varphi_\Gamma(\tau)$ we can conclude that the right sides of Eqs. (37) are stochastically independent for different values of σ. Thus, by calculating the sample correlation function $\chi(\sigma, \tau_0)$ from the signal we obtain a system of linear equations (37) for estimating the coefficients $a_k(\mathbf{x})$ ($k = 0, 1, 2, \ldots, q-1$); the rights sides of the equations in this system contain stochastically independent random quantities with zero mathematical expectations. These conditions correspond to the requirements of the method of least squares; accordingly, we construct the functional

$$\Phi^{(\mathbf{x})}[\alpha_0(\mathbf{x}), \ldots, \alpha_{q-1}(\mathbf{x})] \equiv \sum_{\sigma=1}^{N}\frac{\left\{\left[\frac{d^q}{d\tau^q} + \sum_{k=0}^{q-1}\alpha_k(\mathbf{x})\frac{d^k}{d\tau^k}\right]\chi_\Gamma^{(0;\,\mathbf{x})}(\sigma\tau_0)\right\}^2}{g_\sigma^{(\mathbf{x})}[\alpha_0(\mathbf{x}), \ldots, \alpha_{q-1}(\mathbf{x})]}, \tag{38}$$

where

$$g_\sigma^{(\mathbf{x})} \equiv \frac{1}{\Gamma}\sum_{\gamma=1}^{\Gamma}\left\{\left[\frac{d^q}{d\tau^q} + \sum_{k=0}^{q-1}\alpha_k(\mathbf{x})\frac{d^k}{d\tau^k}\right]\chi_{(\gamma)}^{(0;\,\mathbf{x})}(\sigma\tau_0)\right\}^2$$

are the estimates of the dispersion of the "error" $\varphi_\Gamma^{(0;\,\mathbf{x})}(\sigma\tau_0)$. The values $\hat{\alpha}_0(\mathbf{x}), \hat{\alpha}_1(\mathbf{x}), \ldots, \hat{\alpha}_{q-1}(\mathbf{x})$, providing estimates of the coefficients $\{a_k(\mathbf{x}); k = 0, 1, \ldots, q-1\}$ realize the minimum for $\Phi_{(\alpha_k)}^{(\mathbf{x})}$. As before, in addition to the direct solution of nonlinear equation (38), we can find these values by the successive linear approximations

$$\sum_{\sigma=1}^{N} \frac{1}{g_{\sigma\mu-1}} \frac{d^k}{d\tau^k} [\chi(\sigma\tau_0)] \left\{ \frac{d^q}{d\tau^q} [\chi(\sigma\tau_0)] + \sum_{m=0}^{q-1} \hat{\alpha}_{m\mu} \frac{d^m}{d\tau^m} [\chi(\sigma\tau_0)] \right\} = 0,$$

$$g_{\sigma\mu-1} \equiv \frac{1}{\Gamma} \sum_{\gamma=1}^{\Gamma} \left\{ \frac{d^q}{d\tau^q} [\chi_{(\gamma)}(\sigma\tau_0)] + \sum_{m=0}^{q-1} \hat{\alpha}_{m\mu-1} \frac{d^m}{d\tau^m} [\chi_{(\gamma)}(\sigma\tau_0)] \right\}^2$$

$$(k = 0, 1, \ldots, q-1; \; \sigma = 1, 2, \ldots, N; \; \mu = 1, 2, \ldots); \tag{39}$$

$$\frac{d^q}{d\tau^q} [\chi(\sigma\tau_0)] + \sum_{m=0}^{q-1} \hat{\alpha}_{m0} \frac{d^m}{d\tau^m} [\chi(\sigma\tau_0)] = 0 \quad (\sigma = 1, 2, \ldots, q). \tag{40}$$

Note that system (40) which we have used here to calculate the zeroth approximation of the estimate $\hat{\alpha}_{m0}(x)$, $(m = 0, 1, \ldots, q-1)$, is similar to Eq. (1), written for the dynamic study of a signal from the same object in the tradiational black-box scheme. It is natural to have this correspondence since in the linear case the differential equation describing the correlation function in terms of the shift τ for $\tau \geq \tau_0$, coincides with the equation for the mathematical expectation of the signal (when the time t is replaced by the shift). Allowing for the fact that τ is small in comparison with the relaxation time of the dynamic system of the object and allowing for the condition

$$\frac{d^{2k+1}}{d\tau^{2k+1}} \langle \chi(0) \rangle = 0, \quad \frac{d^{2k}}{d\tau^{2k}} \langle \chi(0) \rangle = (-1)^k \left\langle \left(\frac{d^k U}{dt^k} \right)^2 \right\rangle \quad (k = 0, 1, \ldots), \tag{41}$$

we can say because the fact that the internal noise F(t) is wide band we have a complete set of internal excitations of the initial deflection (41) in the correlation function of the signal from the object.

The above-discussed scheme for analyzing the dynamic structure of an autooscillating object in the neighborhood of the ω-limit cycle is much more complex than a similar method for studying the dynamic characteristics of an object moving in almost stable equilibrium. In fact, not only are Eqs. (38) or (39), (40) simpler than the corresponding formulas [for example, (35)] but there is a great deal of effort involved in finding the position of the mean statistical trajectory and then shifting from Cartesian coordinates in phase space $(U, \frac{dU}{dt}, \ldots, \frac{d^{q-1}U}{dt^{q-1}})$ to local coordinates associated with the cycle [the phase d' and orthogonal deviations n_m (m = 1, 2, ..., q − 1) from the cycle] in order to record the signal, particularly for objects with a large number of degrees of freedom. At the same time, when analyzing the structure of an object in terms of the steady state signal, it is very important to consider motion near the cycle; therefore, we should consider the possibility of considerably simplifying calculation when studying a broad class of autooscillators.

We first note that for the mean statistical signal $U_0^t(t + \tau_x; \mathbf{x})$ of an autonomous oscillator it is not possible to divide the signal recording into identical segments Θ_x by completely copying the analysis procedure in the neighborhood of the stationary point; here Θ_x is the mean period estimated over the entire recording; as a result, we cannot write

$$U_\gamma^{(\mathbf{x})}(t) \equiv U(t_\gamma + t; \mathbf{x}), \quad t_\gamma \equiv -\frac{\Theta}{2} + (\gamma - 1)\Theta_\mathbf{x}, \quad 0 \leqslant t < \Theta_\mathbf{x},$$

$$U_0^+(t + \tau_\mathbf{x}; \mathbf{x}) \sim \frac{1}{\Gamma_\mathbf{x}} \sum_{\gamma=1}^{\Gamma_\mathbf{x}} U_\gamma^{(\mathbf{x})}(t), \quad \Gamma_\mathbf{x} \leqslant \frac{\Theta}{\Theta_\mathbf{x}} < \Gamma_\mathbf{x} + 1 \tag{42}$$

if the autonomous dynamic system near the asymptotically stable limit cycle possesses a qualitatively different reaction to excitations directed across and along the cycle. Whereas deviations in the image point orthogonal to the cycle disappear within a time on the order of $\{\theta_g(x)\}$, the autonomous system has the same reaction to any displacement along the cycle, i.e., it is always in dynamic motion. This leads to the fact that dispersion in the orthogonal deviations of the image point of the object from the cycle proves to be unlimitedly small over an arbitrarily large time interval and sufficiently small internal fluctuations $\langle [F(t)]^2 \rangle$; however, the dispersion in the tangential coordinate (phase) increases unlimitedly with an increase in time: $\langle [t'(t) - t]^2 \rangle \approx \langle [\sigma(0)]^2 \rangle + 2Dt$. Therefore, when the observation time Θ is fairly long, the principal value of the phase

$$t_\mathbf{x}^*(t) \equiv t_\mathbf{x}'(t) - k\Theta_\mathbf{x}, \qquad 0 < t_\mathbf{x}^* < \Theta_\mathbf{x}$$

can be assumed to be distributed uniformly over the interval $0 < t_\mathbf{x}^* < \Theta_\mathbf{x}$; correspondingly, the signal $U(t; \mathbf{x})$ is a stationary random process whose mathematical expectation for fixed $\mathbf{x}$ remains constant, i.e., despite (42) we have

$$\frac{1}{\Gamma_\mathbf{x}} \sum_{\gamma=1}^{\Gamma_\mathbf{x}} U_\gamma^{(\mathbf{x})}(t) \approx \frac{1}{\Theta} \int_{-\frac{\Theta}{2}}^{\frac{\Theta}{2}} U(t; \mathbf{x})\, dt.$$

In the general case of analyzing the dynamic structure in the neighborhood of a nondegenerating ω-limit trajectory, this makes it necessary to carry out the complicated procedure of finding the mean statistical trajectory by successive orthogonal projections in q-dimensional phase space $R_q(\mathbf{x})$, then shifting to the local coordinates t', n_m (m = 1, 2, ..., q−1), calculating the dynamic characteristics in these coordinates, and very often, changing the characteristics back into Cartesian coordinates in space $R_q(\mathbf{x})$. However, for the diffusion coefficient of the autooscillator, we have

$$T_\mathbf{x} \gg 2D(\mathbf{x}) > \frac{T_\mathbf{x}}{\Gamma_\mathbf{x}}$$

for a low internal-fluctuation intensity and a fairly large observation interval. Here, the dispersion in phase diffusion within one period is small in comparison with the square of the period* and we can considerably simplify the above general procedure.

We draw the level line $U \equiv c(\mathbf{x})$ along the signal recording $U(t; \mathbf{x})$ so that on each interval of length $\sim T_\mathbf{x}$ this line intersects the curve $U(t; \mathbf{x})$ at least two times. We let $t_{kc}^{(\mathbf{x})}$

(k = 1, 2, ..., $K_{c\mathbf{x}}$) be the moments in time corresponding to all points of a given signal level: $U(t_k^{(\mathbf{x})}; \mathbf{x}) \equiv c(\mathbf{x})$. For a given position in phase space and a low internal-fluctuation intensity, these points form two or more groups, each of which contains about $\Gamma_\mathbf{x}$ points. By choosing one such group we can find its centroid: $-U_{c,\mathbf{x}}$. We isolate the moments $t_{1,\mathbf{x}}, t_{2,\mathbf{x}}, ..., t_{\Gamma_\mathbf{x};\mathbf{x}}$, separated by intervals of duration $\sim \Theta_\mathbf{x}$, in which the observed motion $U(t; \mathbf{x})$ passes a minimum distance away from the point $U_{c,\mathbf{x}}$ in phase space $R_q(\mathbf{x})$. We define the phase $t_\mathbf{x}'(t)$ of motion at time t letting $t_\mathbf{x}'(t_{\gamma;\mathbf{x}}) \equiv \gamma\Theta_\mathbf{x}$ ($\gamma = 1, 2, ..., \Gamma_\mathbf{x}$), and assuming $t_\mathbf{x}'(t)$ is a function of

---

*The very similar assumption that the random phase of the observed signal is a monotonic function of time should have been mentioned earlier; to be proper, we should make the substitution for the argument $t \to t'(t)$.

given smoothness at intermediate points. Then performing the corresponding local changes in the time scale

$$W(t'_\mathbf{x}; \mathbf{x}) \equiv U(t(t'_\mathbf{x}); \mathbf{x}),$$

we now obtain

$$W_0^+(\theta; \mathbf{x}) \equiv \frac{1}{\Gamma_\mathbf{x}} \sum_{\gamma=1}^{\Gamma_\mathbf{x}} W(\gamma \vartheta_\mathbf{x} + \theta; \mathbf{x}), \xi_\gamma(\theta; \mathbf{x}) \equiv W(\gamma \vartheta_\mathbf{x} + \theta; \mathbf{x}) - W_0^+(\theta; \mathbf{x}), \ 0 \leqslant \theta < T_\mathbf{x};$$

$$\frac{d^q}{d\theta^q}[\xi_\gamma(\theta; \mathbf{x})] + \sum_{k=0}^{q-1} a_k^{(\theta; \mathbf{x})} \frac{d^k}{d\theta^k}[\xi_\gamma(\theta, \mathbf{x})] = F_\gamma(\theta; \mathbf{x}).$$

As for the case of the stable-equilibrium neighborhood the sample correlation functions

$$\chi_\Gamma^{(\theta; \mathbf{x})}(\tau) \equiv \frac{1}{\Gamma_\mathbf{x}} \sum_{\gamma=1}^{\Gamma_\mathbf{x}} \xi_\gamma(\theta; \mathbf{x}) \xi_\gamma(\theta + \tau; \mathbf{x}) \text{ and } \varphi_\Gamma^{(\theta; \mathbf{x})}(\tau) \equiv \frac{1}{\Gamma_\mathbf{x}} \sum_{\gamma=1}^{\Gamma_\mathbf{x}} \xi_\gamma(\theta; \mathbf{x}) F_\gamma(\theta + \tau; \mathbf{x})$$

for motion which is almost dynamically stable and periodic are associated by the equation

$$\left[\frac{\partial^q}{\partial \tau^q} + \sum_{k=0}^{q-1} a_k^{(\theta+\tau; \mathbf{x})} \frac{\partial^k}{\partial \tau^k}\right] \chi_\Gamma^{(\theta; \mathbf{x})}(\tau) = \varphi_\Gamma^{(\theta; \mathbf{x})}(\tau). \tag{43}$$

As before, we can say that the random processes $\chi_{\Gamma(\tau)}^{\theta;x}$ and $\varphi_{\Gamma(\tau)}^{\theta;x}$ are asymptotically normal (for $\Gamma_\mathbf{x} \to \infty$) where in view of (11),

$$\langle \varphi_\Gamma^{(\theta; \mathbf{x})}(\tau) \rangle = 0 \text{ for } \tau > \tau_0,$$

and

$$\langle \varphi_\Gamma^{(\theta; \mathbf{x})}(\tau') \varphi_\Gamma^{(\theta; \mathbf{x})}(\tau'') \rangle \approx 0 \text{ for } |\tau' - \tau''| \geqslant \tau_0.$$

Taking this into account and the fact that the functions $a_\varkappa^{(\theta; \mathbf{x})}$ are periodic in θ:

$$a_k^{(\theta; \mathbf{x})} \equiv \sum_{r=-R}^{R} a_{k,r}^{(\mathbf{x})} \exp\left(i \frac{2\pi r \theta}{T_\mathbf{x}}\right). \tag{44}$$

We use, as before, expressions (43) in order to form the functional $\Phi[\alpha_{k,r}^{(\mathbf{x})}$ $(k = 0, 1, \ldots, q-1;$ $r = 0, \pm 1, \ldots, \pm R)]$. The values of $\hat{\alpha}_{k,r}^{(\mathbf{x})}$ minimizing this functional provide estimates for the corresponding Fourier coefficients $a_{k,r}^{(\mathbf{x})}$ of the expansion in (44).

§4. Region of Application of Method

Thus, for fairly small internal fluctuations when we can ignore nonlinear deviations in the signal from the steady-state dynamic value, the second stage involved in analyzing the object — isolating the corresponding operator from the class $\{D_\alpha\}$ — actually reduces to an autoregression problem involving the method of least squares. The above discussion does not include certain complicating factors. First, in addition to the internal fluctuations of a signal [the response of the dynamic system of the object to the short-correlation excitation F(t)), any recording of the signal contains external noise for the object, i.e., errors arising from the time the signal is generated by the object until this signal is recorded. For a considerable degree of external noise, the errors in estimating these parameters depend not only upon the sample value [the relationship between the observation duration and the correlation interval τ_0 of the perturbation F(t)] and the structure of the dynamic system, but also upon the relationship between the intensity of the external noise and the internal fluctuations [7].

Another complication which is of importance in practice is associated with the fact that the order of the differential dynamic operator is usually unknown and must be estimated. This is a very complicated process, particularly if the observation interval is fairly large and the role of the nonlinear dynamic terms for a given level of internal fluctuations has not been precisely established.

In conclusions we will make a few comments (using very simple examples as illustrations) on the fact that the technique discussed here for obtaining a solution in the second stage of object analysis permits a broadening in the set of classes $\{D_\alpha\}$ (which are otherwise fairly narrow) by *a priori* information.

We first consider a scheme for estimating the parameters of the nonlinear dynamic operator whose general structure is already known. Let the structure of a q-th order ordinary differential operator near the asymptotically stable stationary point $U = U_0$ be such that for a given fluctuation level considerable error is introduced by not allowing for quadratic and cubic terms; in addition, we assume that there is justification for discarding higher powers of the signal fluctuation. Because of nonlinear effects, the mean statistical position of the image point

$$(\langle U(t) \rangle \equiv U_0', \frac{d^k}{dt^k} \langle U(t) \rangle = 0, \ k = 1, 2, \ldots)$$

is displaced from the dynamic stationary point; the degree of displacement depends upon the intensity of the Gaussian fluctuations F(t), and also upon the dynamic structure of the object estimated in the analysis. Without going into calculation, for simplicity, we shall set $U_0' = 0$. Then the signal is described by an equation with p unknown parameters $\{a_k, a_{k,l}, a_{k,l,m}\}$

$$\frac{d^q U}{dt^q} + \sum_{k=1}^{q-1} a_k \frac{d^k U}{dt^k} + \sum_{k=1}^{q-1} \sum_{l=1}^{k} a_{k,l} \frac{d^k U}{dt^k} \frac{d^l U}{dt^l} + \sum_{k=1}^{q-1} \sum_{l=1}^{k} \sum_{m=1}^{l} a_{k,l,m} \frac{d^k U}{dt^n} \frac{d^l U}{dt^l} \frac{d^m U}{dt^m} = F.$$

We shall let

$$U_\gamma(t) \equiv U(t - t_\gamma), \ F_\gamma(t) \equiv F(t - t_\gamma) \ \text{for} \ t_{\gamma-1} \leqslant t < t_\gamma,$$

where

$$t_\gamma = -\frac{\Theta}{2} + (\gamma - 1) S \tau_0; \ \gamma = 1, 2, \ldots, \Gamma + 1; \ \Gamma \gg 1,$$
$$S \gg p, \ \Gamma S \tau_0 = \Theta_0.$$

This corresponds to the discussion given above when analyzing the structure of the neighborhood of the stationary point.

In addition, we define the sample correlation functions by the formulas

$$\xi_\Gamma^{(k)}(s) \equiv \frac{1}{\Gamma} \sum_{\gamma=1}^{\Gamma} \frac{d^k U_\gamma}{dt^k}(t_\gamma) U_\gamma(t_\gamma - s\tau_0), \quad \xi_\Gamma^{(k,l)}(s) \equiv \frac{1}{\Gamma} \sum_{\gamma=1}^{\Gamma} \frac{d^k U_\gamma}{dt^k}(t_\gamma) \frac{d^l U_\gamma}{dt^l}(t_\gamma) U_\gamma(t_\gamma - s\tau_0),$$

$$\eta_\Gamma^{k,l,m}(s) \equiv \frac{1}{\Gamma} \sum_{\gamma=1}^{\Gamma} \frac{d^k U_\gamma}{dt^k}(t_\gamma) \frac{d^l U_\gamma}{dt^l}(t_\gamma) \frac{d^m U_\gamma}{dt^m}(t_\gamma) U_\gamma(t_\gamma - s\tau_0),$$

$$\varphi_\Gamma(s) \equiv \frac{1}{\Gamma} \sum_{\gamma=1}^{\Gamma} F(t_\gamma) U(t_\gamma - s\tau_0) \quad (s = 1, 2, \ldots, S).$$

As before, it is clear that

$$\xi_\Gamma^{(q)}(s) + \sum_{k=0}^{q-1} a_k \xi_\Gamma^{(k)}(s) + \sum_{k=0}^{q-1} \sum_{l=0}^{k} a_{kl} \xi_\Gamma^{(k,l)}(s) + \sum_{k=0}^{q-1} \sum_{l=0}^{k} \sum_{m=0}^{l} a_{klm} \xi_\Gamma^{(k,l,m)}(s) = \varphi_\Gamma(s). \tag{45}$$

As in the considered variant where the equation is linear with respect to the signal, all sample correlation functions ξ and φ are also asymptotically normal (for $\Gamma \to \infty$); moreover, we again have

$$\langle \varphi_\Gamma(s) \rangle = 0, \quad \langle [\varphi_\Gamma(s)]^2 \rangle = \frac{1}{\Gamma} \langle [F(t_\gamma)]^2 [U(t_\gamma - s\tau_0)]^2 \rangle \quad (s = 1, 2, \ldots, S).$$

However, it is now impossible to assume the signal fluctuations are Gaussian. It is therefore not possible in the general case to state beforehand that the correlation function (and thus the stochastic relationship) of the values $\varphi_\Gamma(s_1)$ and $\varphi_\Gamma(s_1)$ disappears for $s_1 \neq s$. The autocorrelation duration of the fluctuations $\varphi_\Gamma(s)$ should now be verified; in the general case, this is done by direct estimates from the signal recording. After this is done, the ordinary method of least squares can be applied to Eq. (45) which is linear in the estimated parameters $\{a_k, a_{k,l}, a_{k,l,m}\}$.

A condition more basic than the one stating that the response of the dynamic system to a perturbation F(t) is linear is the assumption that the observed changes in the signal do not cause the image point to leave the neighborhood of the asymptotically stable dynamic trajectory. Obviously, this assumption usually corresponds to the conditions of observing the steady-state signal when the dynamic operator has asymptotically stable trajectories. It is clear that under the effect of the fluctuation perturbation the image point of the object will not remain for any great length of time in the small neighborhood of neutral equilibrium (this is characterized by the time-increase in dispersion of the deviations) and therefore in the neighborhood of the unstable trajectory (associated with a nonzero regular component for the deviation velocity). The situation can change if we allow for the dependence of the motion on the parameters. We shall briefly consider the case when the quantities $\zeta_1, \zeta_2, \ldots, \zeta_Q$ which vary with time, considerably complicate the behavior of the object; however, in certain regions where important data can be obtained on phenomena of interest to the researcher, the following parameters of the problem are constant and can be assumed to be fixed: $\zeta_k(t) \equiv x_k$, $t_{\gamma_1} \leqslant t < t_{\gamma_2}$ ($\gamma = 1, 2, \ldots, \Gamma$; $k = 1, 2, \ldots, Q$). Here, it is convenient to analyze the signal on the corresponding sections of the observation interval only and to write

$$\frac{d^q U}{dt^q} + a\left(U, \frac{dU}{dt}, \ldots, \frac{d^{q-1}U}{dt^{q-1}}; \mathbf{x}\right) \equiv F(t; \mathbf{x}),$$

$$t_{\gamma,1} < t < t_{\gamma,2} (\gamma = 1, 2, \ldots, \Gamma), \quad \mathbf{x} \equiv (x_1, x_2, \ldots, x_Q).$$

The initial values $\frac{d^k U}{dt^k}(t_{\gamma 1}; \mathbf{x}) \equiv U_\gamma^{(k)}(\mathbf{x}) (k = 0, 1, \ldots, q-1)$ can have very broad distributions; this leads to the fact that the positions of the object impage points in phase space $R_q(\mathbf{x})$ which are observed for $t_{\gamma,1} < t < t_{\gamma,2} (\gamma = 1, 2, \ldots, \Gamma)$ do not contract to a one-dimensional or zero-dimensional trajectory, but "spread out" into a broad q-dimensional region. Here, the requirements imposed on the duration and detail of signal recording become much more critical and the amount of difficulty involved in numerically processing the data increases considerably. It is now not possible to make use of the fact that the material is uniform by expanding the signal and its derivatives in powers of the deviations from the ω-limit trajectory, calculating the coefficients on this trajectory, and discarding terms with a high order of smallness. The situation improves when there is such a wealth of *a priori* information on the observed processes that the problem involved in the second stage of analyzing the object reduces to an estimate of a few parameters.

We shall give an illustration on the above statement that the difficulties involved in analyzing the direct problem are not repeated for inverse problems. We begin with the situation in which the dynamics of the observed steady-state signal are described by a very simple differential-difference equation. We assume that the known signal from the object satisfies the equation

$$\frac{dU}{dt}(t) + aU(t - \Delta) = F(t),$$

for given observation conditions where $F(t)$ are the short-correlation fluctuations; here it is necessary to estimate the parameters a and Δ. When $\Delta \gg \tau_0$ we have

$$a = -\frac{\left\langle U(\theta) \frac{dU}{dt}(t) \right\rangle}{\langle U(\theta) U(t - \Delta) \rangle} = \text{const}, \quad \theta - t \geqslant \tau_0. \tag{46}$$

It is natural to assume the random process $U(t)$ is stationary; here

$$\langle U(\theta) U(t - \Delta) \rangle = \chi(\tau - \Delta), \quad \left\langle U(\theta) \frac{dU}{dt}(t) \right\rangle = \frac{d}{d\tau} \chi(\tau), \quad \tau \equiv t - \theta.$$

These considerations make it possible to estimate the delay Δ. By estimating the delay we can use (46) to find the coefficient a.

We shall now consider a method for analyzing the observation results for a two-dimensional signal (U_1, U_2) described by a dynamic partial-differential equation where one of the arguments is the time t. Let it be necessary to estimate the parameters a_1, a_2, $b_{1,1}$, $b_{1,2}$, $b_{2,1}$, $b_{2,2}$ of first-order partial differential nonlinear operators where the equations of the signal have the form

$$\partial U_1(t, x)/\partial t + a_1 \partial U_1(t, x)/\partial x + b_{1,1} U_1(t, x) + b_{1,2} U_2(t, x) = F_1(t, x),$$
$$\partial U_2(t, x)/\partial t + a_2 \partial U_2(t, x)/\partial x + b_{2,1} U_1(t, x) + b_{2,2} U_2(t, x) = F_2(t, x), \tag{47}$$

and the short-correlation functions $F_1(t, x)$ and $F_2(t', x)$ are stochastically independent for all t and t'. We shall assume that observations of the signal components U_1 and U_2 are taken over time intervals

$$t_{\gamma, 1} < t \leqslant t_{\gamma, 2}, \quad t_{\gamma, 2} = t_{\gamma, 1} + \Theta_1, \quad t_{\gamma, 2} < t_{\gamma+1, 1},$$

$$\frac{\Theta_1}{6\tau_0} > 1, \quad \gamma = 1, 2, \ldots, \Gamma, \quad \Gamma \gg 1 \tag{47}$$

and discrete values of the second argument $x = x_1, x_2, \ldots, x_P$; $P \gg 1$, where (for clarity) $x_p - x_{p-1} = 2^{-p}$ ($p = 1, 2, \ldots, P$). Then, as p increases, the quantities $\frac{\partial U_i}{\partial x}(t, x_p)$ provide better and better estimates of the partial derivatives $2^p[U_i(t, x_p) - U_i(t, x_{p-1})]$. Allowing for this fact, we introduce the parameters $a_i^{(p)}$, $b_i^{(p)}$, and write the system of equations

$$\frac{\partial U_{i\gamma}}{\partial t}(t, x_p) + a_i^{(p)} 2^{(p)} [U_{i\gamma}(t, x_p) - U_{i\gamma}(t, x_{p-1})] +$$
$$+ b_{i1}^{(p)} U_{1\gamma}(t, x_p) + b_{i2}^{(p)} U_{2\gamma}(t, x_p) = F_{i\gamma}(t, x_p)$$
$$(i = 1, 2; \; \gamma = 1, 2, \ldots, \Gamma; \; p = 1, 2, \ldots, P), \tag{48}$$

which is then analyzed in detail in the same manner as a fluctuation-perturbation system of ordinary differential equations. We introduce the following notation for the same correlation functions:

$$\frac{1}{\Gamma} \sum_{\gamma=1}^{\Gamma} U_{i\gamma}(t_{\gamma_2} - s\tau_0, x_p) \frac{\partial U_{j\gamma}}{\partial t}(t_{\gamma_2}, x_p) \equiv \Xi_{ij}(s, p),$$

$$\frac{1}{\Gamma} \sum_{\gamma=1}^{\Gamma} U_{i\gamma}(t_{\gamma_2} - s\tau_0, x_p) U_{j\gamma}(t_{\gamma_2}, x_p) \equiv \xi_{ij}(s, p),$$

$$\frac{1}{\Gamma} \sum_{\gamma=1}^{\Gamma} U_{i\gamma}(t_{\gamma_2} - s\tau_0, x_p) U_{l\gamma}(t_{\gamma_2}, x_{p-1}) \equiv \eta_{ji}(s, p),$$

$$\frac{1}{\Gamma} \sum_{\gamma=1}^{\Gamma} U_{i\gamma}(t_{\gamma_2} - s\tau_0, x_p) F_{j\gamma}(t_{\gamma_2}, x_p) \equiv \varphi_{ij}(s, p).$$

According to (48), these functions are associated by the equations.

$$\Xi_{ij}(s, p) + a_j^{(p)} 2^p [\xi_{ij}(s, p) - \eta_{lj}(s, p)] + b_{j1}^{(p)} \xi_{i1}(s, p) + b_{j2}^{(p)} \xi_{i2}(s, p) = \varphi_{ij}(s, p)$$

$$(i, j = 1, 2; \; s = 1, 2, \ldots, S; \; p = 1, 2, \ldots, P). \tag{49}$$

The right sides $\varphi_{ij}(s, p)$ of these equations are fluctuating and the dispersions, which approach zero as Γ increases, are not correlated for different s. As before, we construct the functional

$$\Phi_p(\alpha_i^{(p)}, \beta_{i,j}^{(p)}; \; i, j = 1, 2) \equiv \sum_{s=1}^{S} \sum_{i, j=1}^{2} \frac{[\varphi_{i,j}^{(s, p)}(\alpha_i^{(p)}, \beta_{ij}^{(p)}; \; i, j = 1, 2)]^2}{g_{i,j}^{(s, p)}(\alpha_i^{(p)}, \beta_{i,j}^{(p)}; \; i, j = 1, 2)}, \tag{50}$$

where, according to (48) and (49),

$$\varphi_{ij}^{(s,p)} \equiv \Xi_{ij}(s,p) + \alpha_j^{(p)} 2^{(p)} [\xi_{ij}(s,p) - \eta_{ij}(s,p)] + \beta_{j1}^{(p)} \xi_{i1}(s,p) + \beta_{j2}^{(p)} \xi_{j2}(s,p),$$

$$g_{ij}^{(s,p)} \equiv \frac{1}{\Gamma} \sum_{\gamma=1}^{\Gamma} [U_{i\gamma}(t_{\gamma 2} - s\tau_0, x_p)]^2 \left\{ \frac{\partial U_{j\gamma}}{\partial t}(t_{\gamma 2}, x_p) + \right.$$

$$\left. + \alpha_j^{(p)} 2^p [U_{j\gamma}(t_{\gamma 2}, x_p) - U_{j\gamma}(t_{\gamma 2}, x_{p-1})] + \beta_{j1}^{(p)} U_{1\gamma}(t_{\gamma 2}, x_p) + \beta_{j2}^{(p)} U_{2\gamma}(t_{\gamma 2}, x_p) \right\}^2.$$

If *a priori* assumptions on the structure of dynamic operator (47) are valid and the observation conditions for the signal satisfy the requirements in (47') where the external noise is fairly low, the values $\hat{\alpha}_i^{(p)}$, $\hat{\beta}^{(p)}_{ij}$ (i, j = 1, 2), for which the functional (50) is a minimum must become almost independent of the index p as p increases. Then, we assume that the quantities $\hat{\alpha}_i^{(p)}$ provide estimates of the coefficient a_i and $\beta_{i,j}^{(p)}$ provide estimates of $b_{i,j}$.

Clearly, within the framework of the inverse analysis problem we can also consider equations which do not have unique solutions. Thus, if the changes in the signal are described by the system of equations $D_i[U(t)] = F_i(t)$ (i = 1, 2, ..., I), it is quite permissible to introduce an estimate for the parameters of any set of equations in this system of interest to the researcher without knowing the structure of the remaining equations. For example, according to *a priori* information, let the observed signal satisfy the equation

$$\frac{\partial U}{\partial t}(t,x) + \sum_{k=0}^{K} a_k \int_{x'}^{x''} \xi^k U(t,\xi) d\xi = F(t,x),$$

where F(t, x) are short-correlation fluctuations and the numbers $\{a_k\}$ must be estimated from the recording of the signal. If the signal U(t, x) is recorded on time intervals (47') for discrete values of the argument (x = x_m):

$$x_m = x' + m\Delta, \quad (m = 0, 1, \ldots, M), \quad M\Delta = x'' - x', \quad M \gg 1,$$

then, letting for example,

$$\int_{x'}^{x''} \xi^k U(t,\xi) d\xi \cong \Delta \sum_{m=1}^{M} (x' + m\Delta)^k U(t, x' + m\Delta),$$

$$\frac{1}{\Gamma} \sum_{\gamma=1}^{\Gamma} U_\gamma(t_{\gamma 2} - s\tau_0, x' + l\Delta) \frac{\partial U_\gamma}{\partial t}(t_{\gamma 2}, x' + n\Delta) \equiv \Xi_{ln}(s),$$

$$\frac{1}{\Gamma} \sum_{\gamma=1}^{\Gamma} U_\gamma(t_{\gamma 2} - s\tau_0, x' + l\Delta) U_\gamma(t_{\gamma 2}, x' + m\Delta) \equiv \xi_{lm}(s),$$

$$\frac{1}{\Gamma} \sum_{\gamma=1}^{\Gamma} U_\gamma(t_{\gamma 2} - s\tau_0, x' + l\Delta) F_\gamma(t_{\gamma 2}, x' + n\Delta) \equiv \varphi_{ln}(s)$$

we have, as before, a system of equations associating the sample correlation function Ξ, ξ obtained from the recording of the signal with the estimated parameters a and the sample correlation functions which, on the average, become zero and lose the correlation for shift s:

$$\Xi_{ln}(s) + \Delta \sum_{k=0}^{K} a_k \sum_{m=1}^{M} (x' + m\Delta)^k \xi_{lm}(s) = \varphi_{ln}(s)$$
$$(l, n = 0, 1, \ldots, M; \; s = 1, 2, \ldots, S).$$

These functions can be calculated directly from the signal. We will not write again the functional which, upon minimization, yields estimates of the parameters a_k.

§5. Comments

A fairly general and simple principle for recording a signal has been formulated and, to some degree, justified. According to this fundamental thesis, the signal properties which are quantitatively important and observed regularly under certain conditions, are associated by some dynamic structure of the object. The role of the object movements which are less important under these conditions (as is the role of the external medium) reflects in this description the time-fluctuating* force F(t) exciting the dynamic system. As shown above, a study of the statistical properties of the response of a dynamic system to a fluctuating perturbation F(t) makes it possible to estimate the dynamic characteristics of an unregulatable object from the steady-state signal for a wide range of problems. The main purpose of this study was not to develop a rigid justification of optimum mathematical calculation schemes for any particular situation but to demonstrate the uniqueness and effectiveness of the formulated principle as a starting point for various natural-science studies. Its systematic utilization (although after many time-consuming operations) makes it possible to obtain material which is organized for further logical analysis. By numerically processing the signal using the black-box scheme without input, the researcher can obtain original and significant prepared data even for problems being worked on intensely or already worked out (see, for example, [9-11]). Unfortunately, despite the relatively short time which has passed since this analysis technique has been considered, erroneous studies have already been performed in astrophysics which have been initiated by this discussion (see, in particular, [8]). The appearance of these studies has provided an important stimulus for the writing of this article which contains a detailed critique of the starting point for the black-box scheme without input.

LITERATURE CITED

1. L. I. Gudzenko, Radiofizika, Vol. 5, No. 3 (1962).
2. M. Arato, A. N. Kolmogorov, and Ya. G. Sinai, Dokl. Akad. Nauk SSSR, Vol. 146, No. 4 (1962).
3. E. Hennan, Analysis of Time Series [Russian translation], Nauka (1964).
4. L. I. Gudzenko and V. E. Chertoprud, Radiofizika, Vol. 8, No. 6 (1965).
5. V. V. Nemytskii and V. V. Stepanov, Qualitative Theory of Differential Equations [in Russian], Moscow, Gostekhizdat (1949).
6. A. N. Kolmogorov, "Justification of the method of least squares," Usp. Mat. Nauk, Vol. 1, No. 1 (11) (1946).
7. L. I. Gudzenko and V. E. Chertoprud, Radiofizika, Vol. 10, No. 3 (1967).
8. N. E. Kurochkii, Astronomicheskii tsirkulyar, No. 429 (1967).
9. L. I. Gudzenko and V. E. Chertoprud, Astron. Zh., Vol. 42, No. 2 (1965).
10. L. I. Gudzenko, Mechanism of Solar Cyclic Activity [in Russian], FIAN Preprint No. 24 (1967).
11. L. M. Ozernoi and V. E. Chertoprud, Astron. Zh., Vol. 44, No. 3 (1967).

---

*We again note the importance to the discussed scheme of analysis of the assumption that it is possible to separate the behavior of an object into dynamic and statistic positions which are functions of the time t since only then can the causality principle be applied. One of the errors in [8] is the result of not understanding this.